Canals and Inland Navigation

Studies in Transport History

Series Editor: John Armstrong

Coastal and Short Sea Shipping
John Armstrong

Railways (volume I)
Terry Gourvish

Railways (volume II)
Geoffrey Channon

Road Transport in the Horse-Drawn Era
Dorian Gerhold

Air Transport
Peter J. Lyth

Canals and Inland Navigation
Gerald Crompton

Motor Transport
Theo Barker

The World of Shipping
David M. Williams

Canals and Inland Navigation

Edited by

GERALD CROMPTON

SCOLAR PRESS

Published by
Scolar Press
Gower House
Croft Road
Aldershot
Hants GU11 3HR
England

Ashgate Publishing Company
Old Post Road
Brookfield
Vermont 05036–9704
USA

British Library Cataloguing in Publication Data
Canals and Inland Navigation
(Studies in Transport History)
1. Canals—Great Britain—History. 2. Inland navigation—
Great Britain—History.
I. Crompton, Gerald.
386.4'0941

ISBN 1–85928–346–2

Library of Congress Cataloging-in-Publication Data
Canals and inland navigation/edited by Gerald Crompton.
p. cm. (Studies in Transport History).
ISBN 1–85928–346–2 (cloth)
1. Canals—Great Britain—History. 2. Inland navigation—Great
Britain—History. 3. Inland navigation—Netherlands—History.
I. Crompton, Gerald. II. Series.
HE395.G74C36 1996
386'.4'0941—dc20 96–22321
CIP

ISBN 1 85928 346 2
Printed on acid free paper
Printed in Great Britain by
Ipswich Book Company, Suffolk

Contents

Studies in Transport History

Series Editor's Preface

The idea for this series originated in a meeting of the Editorial Board of *The Journal of Transport History*. As part of the celebrations to mark the fortieth anniversary of the founding of the journal, a classified index of the articles appearing in each of the series was compiled by Julie Stevenson and published in the September 1993 number of the journal. This exercise revealed the wealth of material that lay in the journal but which was largely forgotten and relatively inaccessible, unless a full run of the journal was close to hand. The index took the first step in making it simpler to determine what was in the journal. The Editorial Board decided it should continue this process by making the best of the essays more readily available.

Hence it is proposed to issue a series of volumes, each containing a number of articles reprinted from *The Journal of Transport History*, dealing with a particular mode of transport. As initially planned there are eight volumes in this series, each covering one form of transport and edited by a different member of the Editorial Board. Each volume contains about ten articles, reprinted from *The Journal of Transport History*. The articles have been selected because they were seminal at the date of their publication and have stood the test of time. Hence they are still important contributions to the state of our knowledge on the topic.

Each volume also contains an introduction by the editor of the volume, which seeks to contextualise the articles selected; that is, to show how each contribution fitted into the state of knowledge at the time of its publication, to explain its importance, to indicate any more-recent literature on the topic that might moderate its findings, and suggest how the debate has moved since the initial publication. Each article retains its original pagination, as in the journal. This will allow citations to be made to the original source.

Thus this series aims to make easily available the most important essays which have appeared in *The Journal of Transport History* over the last forty years, collected together on a thematic basis. It is hoped that this will encourage and provoke further research into transport history, in order to carry the debates into the twenty-first century.

John Armstrong
1996

Thames Valley University

Introduction: rivers, canals and economic development

GERALD CROMPTON

This collection of articles covers a time span of more than 300 years from the beginning of the seventeenth century to the early twentieth. It also explores several distinct themes in the history of inland navigation. These include the institutional obstacles to river improvements in the pre-industrial period; the effects of an artificial cut on an established river navigation (a subject linking with the canal era); an early and rare contribution to the principles of canal management; responses to the engineering problems posed by canal-building in a geographically difficult area; two studies of the living standards of the canal labour force, including an account of an official strike; and an interpretive outline of the development of the transport system in a country where waterways had a high profile. The remaining articles study the origins and impact of a number of canal projects on the local or regional economies of various parts of Britain - South Wales, Lancashire, Yorkshire and the East Midlands.

The first article is concerned exclusively with the pre-canal era. Chalklin discussed three separate attempts during the seventeenth century to extend navigation on the River Medway above Maidstone.[1] He found that the gentry and farmers who owned the river banks were able to thwart all three schemes, being determined to protect their weirs, fences and fish stocks, and to prevent the construction of towpaths or widening of the river. To many others the benefits obtainable from improved river transport were obvious and wide in scope, for both the Medway valley and for Kent as a whole. Foodstuffs and raw materials for the cloth industry would have been more easily brought up river, and timber and iron from the Weald conveyed downstream. The Government had an interest in the supply of the naval dockyards at Chatham, which was no doubt part of the reason for its general sympathy with the supporters of improvement.

The unsuccessful schemes varied in character. In 1600 a major reason for failure was the inadequate powers of the agency in question, the local Commission of Sewers, which was unable to challenge effectively the property rights of the hostile landowners. T. S. Willan's pioneering work of 1936 summed up this episode by noting that the Commissioners had succeeded in asserting their powers over the cleansing of the river, but not over its navigation.[2] Chalklin went on to examine two additional contested projects which had not been covered by Willan's researches. In 1627 and afterwards, a contractor with the support of the Privy Council made some progress in clearing obstacles, but the defeat of a bill in Parliament undermined his chance of ultimate success. In 1665 the situation looked distinctly more promising. An Act was passed to make the Medway navigable, and ten local landowners were appointed undertakers, under the leadership of one who held a large estate of coppice woodland, plus an iron furnace and forge. Despite the involvement of such relevant interests, this third attempt seems to have failed through lack of money. It was not until 1740 that the river was made navigable as far as Tonbridge.

Chalklin's study remains a powerful indicator of the force of the obstacles to improved river navigation right up to the eve of the canal age. His examples were all the more striking in view of the economic importance of the Weald, almost the only part of Kent more than ten or fifteen miles from the sea, and otherwise dependent on poor roads for communication with London and the Thames.[3]

Duckham's article was focused on the stage in the history of the Aire and Calder Navigation between the 1770s and the 1820s when it had a significant impact on the town of Selby.[4] Over a longer period, and more importantly, it had positively influenced the growth of Leeds and Wakefield. The proprietors of the Aire and Calder were responsible for blocking one proposal, for a Leeds-Selby canal, which would undoubtedly have benefited Selby. However, between 1774 and 1778 a five-mile canal was cut from Haddlesey to Selby. This provided an improved route between the old Aire and Calder and the River Ouse, by-passing the difficult tidal section of the Aire. The brief new link connected Selby, with its established coastal trade, to the urban West Riding. When steampackets appeared on the river after 1815, the town became a transfer point from stagecoaches to steamboats (especially to Hull).

Duckham documents convincingly the growth in Selby's prosperity, which was based on a greatly expanded river-borne trade, with considerable development of related manufacturing and service activities. An inherent flaw in this position of apparent strength was the excessive shallowness of the Selby canal (a draught of only three and a half feet), and the fact that culverts beneath it made deepening unviable. The opposition of landowners worried about flooding had been a factor in the adoption of such arrangements in the 1770s, as was a failure to foresee future traffic levels. This chapter in the history of the Aire and Calder was closed with the completion of the Knottingley-Goole canal in 1826, which naturally favoured Goole rather than Selby. For the latter, this implied 'devitalisation', though not 'disaster'.[5]

The article was concerned with a locally significant side-effect of the development of a major inland waterway. The principal purpose and consequence of the Aire and Calder Navigation was the linking of the urban centres of West Yorkshire with the coast. The Selby Canal stimulated this flow, and for half a century was the overwhelmingly dominant influence on the growth of a small commercial town.

The piece by Harris on Merseyside dealt with a crucial region in the early canal-building era.[6] The same author published a large-scale study of one area, St Helens and its locality, within the Merseyside region.[7] This was a joint work, with T. C. Barker, who had in 1960 written about the Sankey Navigation, appropriately, as the 'beginnings of the canal age' in Britain.[8] Harris emphasised the huge advantage derived by the port of Liverpool from its position as a canal terminal, and the early date by which this was achieved. He also stressed the centrality of coal supply in all transport considerations, and the highly supportive attitude of the merchant-dominated Liverpool Corporation. Although the South-West Lancashire coalfield was no more than ten miles away at the nearest point, the growth prospects of a large commercial and industrial city were critically dependent in the second half of the eighteenth century on fundamental improvements in water transport in the hinterland. These circumstances explain the capacity of 'canal controversies' to absorb the attention of a wide community.

The solution to Liverpool's problems came in three main stages. The first two were represented by the Douglas Navigation (1742) and the Sankey Canal (1757). Both made positive contributions and certainly sufficed to prevent deterioration in the city's fuel supply. They were not, however, able to eliminate periodic rises in the price of coal, and hence served to stimulate further interest in canal projects. The idea of a Leeds - Liverpool canal was actively promoted in 1766, with much of the initiative, in contrast with the Sankey, coming from Yorkshire. Harris unravelled the complex processes by which the Leeds and Liverpool finally gained sufficient support at the Merseyside end to win parliamentary sanction. Partly because the proposed route did not, in its initial form, offer adequate connections with the coalfield in the Wigan area, the whole project was for some time vulnerable to a plausible rival scheme, the Liverpool canal. This concept assumed various forms, including a link between the Leeds and Liverpool and the Trent and Mersey via a bridge over the Mersey at Runcorn. Another aspect involved a plan for a direct Liverpool-Wigan canal. The upshot was that the Leeds and Liverpool acquired the Douglas Navigation in 1771, and made convincing arrangements to connect with the coalfield at Wigan. The Liverpool canal, though 'an idea of real merit', remained 'an intriguing might-have-been',[9] defeated by rival vested interests, frequent changes of tack and failure to engage a well-known engineer. The Leeds and Liverpool proceeded to justify itself abundantly as a supplier of coal. The results have since been analysed more fully in Langton's study of the region. This concluded that by 1800 almost the whole of the South-West Lancashire coalfield formed 'a single economic system with integrated patterns of production, prices and competition around the hub of Liverpool'. At that date the city was consuming about 200,000 tons a year and exporting around 150,000 tons.[10]

Firth studied the origins of the Leeds and Liverpool from a different standpoint.[11] He was interested initially in the industrialisation of Bradford, which in the mid-eighteenth century was still a small market town on the north-west fringe of the Yorkshire coalfield. Cheaper transport between Bradford and Skipton was a necessity for the development of the local coal industry, on the basis of an interchange with the limestone (for farming and building) which was abundant in the Craven district on the Lancashire-Yorkshire borders. The notion of a trans-Pennine canal was a further step of substantial dimensions, but one which became more credible in the light of the success of the Sankey and Bridgewater enterprises. Firth was confident that 'the initiative came from within the cabal of Bradford colliery owners.'[12] His own preference was to correct what he regarded as previous underestimations of the promotional role of John Stanhope, a Bradford attorney and landowner with industrial interests. After Stanhope's death, the leading part fell to John Hustler, a local woolstapler with significant Quaker connections. After extended difficulties with the Lancashire supporters of the project, an Act was obtained in 1770. By the late 1770s the canal was open in two key sections - from Liverpool to the River Douglas in Lancashire, and from Gargrave to Leeds in Yorkshire. From 1774 a short branch had linked Bradford to the main canal. Work came to a lengthy halt in the late 1770s, but both parts of this ambitious project flourished, with the Lancashire side enjoying a much larger revenue. The economy of Bradford benefited greatly from the completion of the Yorkshire section. By the 1790s the Low Moor Iron Works had become the first successful large-scale industrial enterprise in the area.

In the opening phases of the Leeds and Liverpool Canal control seems to have rested in Bradford, and especially with Hustler. The proprietors were divided. Initially at least, the Yorkshire interests generally favoured high tolls so as to maximise revenue. Those at the Lancashire end preferred low tolls to promote the coal trade.[13] In terms of the distinction suggested by Ward, the 'financial' motive predominated east of the Pennines, and the 'economic' motive to the west.[14] In apparent contradiction of this pattern, however, Hustler himself acquired coal interests in Lancashire, and became a customer as well as a proprietor of the company. Langton had identified him and Jonathan Blundell as the dominant coalowners of the Orrell/Upholland area and members of a clique which controlled the best reserves of coal, their distribution and the policy of the canal company.[15] Firth's article opened up a new angle on the early history of the Leeds and Liverpool, as Lancashire had previously enjoyed a near monopoly of historical attention. It did not, however, explore the issue of the economic justification of the canal as a trans-Pennine project by evaluating the potential for long-distance through traffic.

Coal likewise held centre stage in the article in which Griffin challenged some existing interpretations of the development of the Leicestershire coalfield in the late-eighteenth and early-nineteenth centuries.[16] Up to the mid-eighteenth century Leicestershire coal had dominated the county market because it was protected by high transport costs against competition from cheaper products to the north and west. From the late 1770s the Erewash Canal and the Soar Navigation opened the market to coal from the Erewash valley and Derbyshire. Plans to extend the Soar Navigation from Loughborough southwards to Leicester were resisted by local coal interests. It was eventually decided as a compromise in 1790 that the Leicester Navigation Company would build two new links - one from Loughborough to Leicester and another from the colliery districts in the west of the county to Loughborough. The latter was to follow the Charnwood or Forest line and was to be accomplished by a combination of canal and tramway. Officially both these routes opened in 1794, but the Forest line experienced such severe technical problems, including the disastrous bursting of a reservoir, that it was never able to function effectively.

Griffin's argument was that the failure of the Forest line had been blamed to an excessive degree for the retarded development of the Leicestershire coalfield. He cited various unrelated reasons to explain setbacks to coal output, and also claimed that the pits in question did in fact stage a vigorous recovery in the early-nineteenth century, well before the railway age. Techniques previously employed by Richard Wilson in a study of the West Riding of Yorkshire (transport dues as an index of economic growth)[17] were used to show that Leicestershire coal was marketed in respectable quantities despite having to use toll roads.

The article is unusual in that it sought to qualify, rather than to illustrate or emphasise, the significance of inland navigation in widening the market for bulk products. A strong position was established in support of the proposition that the fortunes of at least one small coalfield varied largely independently of water transport facilities. Although Griffin presented his thesis as revisionist, the argument was not completely novel. Hadfield had noted in 1966 that the colliery owners realised even before the opening of the Forest line that it would not be helpful to them.[18] It was also highly specific to one group of pits and one locality. The value of the Erewash/Soar route to both Derbyshire producers and Leicestershire consumers was not disputed.

What emerges clearly from the complex subject-matter of Griffin's article (and from several others in this collection) is that transport improvements were never experienced as universally beneficial. Many expected to gain, either directly as investors, (the private rate of return) or indirectly as shippers, purchasers of raw materials, consumers etc. (the social rate of return).[19] But invariably some interests stood to suffer an absolute or relative disadvantage from a new development, because of the competitive implications for established transport facilities, manufacturing companies or urban centres. As Flinn recognised in his volume on the history of the coal industry, 'any canal necessarily threatened to alter the existing market structure'.[20] This was because the purpose of every new one was to enlarge the market for some pits, or to facilitate entry into a given market. On balance, of course, the gainers were far more numerous than the losers, and the social rate of return was nearly always positive even in those cases where the private rate was not.

Pollin's article on the Swansea canal was one of the most effective studies of a single canal and its local impact.[21] It told a story of success sustained over an unusually long period which extended into the railway age. Indeed the emergence of serious competition was so long delayed that the Swansea might in its latter days have been more notable for its private than its social rate of return.

Pollins stressed the importance of the new canal to the economy of Swansea and the Tawe valley in the first half of the nineteenth century, with particular reference to the coal, iron and copper trades. Although parliamentary sanction was obtained during the so-called 'mania' of the early 1790s, the author was able to demonstrate that a very high proportion of investors were local, led by a solicitor and mineral owner. Two main groups were identified among the backers of the canal - those supporting the development of the town and port of Swansea, and those with business interests in the valley who wanted better communications with the port.

The company was a resounding financial success, paying 10 per cent or more in dividends in forty-seven out of sixty-nine years between 1804 and 1872. Fluctuations were experienced, but the long-term record was one of immense strength. Even the arrival of the Llanelly Railway in the early 1840s, which siphoned off part of the canal's traffic, failed to prevent profits from reaching a peak in the 1850s. An important feature of the company's development was that short branches and feeder tramways, normally built by customers, were added at several points to link with collieries and business premises.

A major element in the prolonged success of the Swansea Canal was the fact that the first railway along the valley was not built until 1860-61. Even then considerable traffic was retained, including the long-distance freight of the Ystalyfera iron works, although overall profits declined. Recognising the inevitability of eventual defeat at the hands of the railways, the directors pulled their master-stroke in 1871. The canal was leased, not to its most direct competitor, the Swansea Vale line, but to the rival Great Western (GWR). The terms of this deal were excellent for the canal shareowners. The GWR cumulative preference shares they received virtually guaranteed them an annual return of 10 per cent, although the canal became a financial burden to the railway. The richest slice of the valley's trade went to the Midland Railway which absorbed the Swansea Vale in 1876.

Pollins explicitly concentrated on the history of the canal, and dealt only incidentally with its economic impact. Subsequent work has supplied the framework within which the latter can be more clearly appreciated. The Swansea was one of four substantial canals built in South Wales during the 1790s (the others were the Glamorganshire, the Monmouthshire and the Neath). Their economic function was essentially to boost the two major industries of the region, coal and iron. In the pre-canal era strategic resources of coal, ironstone and limestone were tantalisingly close to the coast, but not sufficiently near for their full potential to be realised. Deep, narrow valleys provided the only obvious conduit for the transfer of heavy, bulky goods from the northern rim of the coalfield to the coastal ports. Hence canal construction on a limited scale (77 miles were built at a cost of less than half a million pounds in the 1790s) made a disproportionately large contribution to economic development.[22] The effect was to promote regional concentration in South Wales, and as Pollard and others have since recognised, this process was a main causal factor in the industrialisation of the British economy.[23] Another historian who emphasised the cumulative effects of the intra-regional benefits induced by canals was Turnbull. His concept of 'a narrow corridor of superior locational space' along the route of a canal is illuminating in the context of South Wales.[24] Geography dictated that the river valleys would always enjoy locational advantages, but canals invested the space with greatly enhanced superiority.

By 1830 South Wales was Britain's major iron-producing region. Although this supremacy was not maintained for long, total output continued to increase rapidly for some decades. Within the region the development of the western area, including Swansea, was outstripped by that of Cardiff and Newport. Here, again, transport factors were relevant. The western area was less well served by railways. The progress of Swansea was further hampered by the inability of its corporation, which opposed a new Dock Bill in 1847, to decide whether it preferred a future as an industrial centre or as a fashionable watering place.[25]

Davies, in 1980, produced one of the few studies which focused on issues of canal management.[26] This study carried additional interest as it was based on ideas formulated by the pottery manufacturer Josiah Wedgwood, who had already been identified as one of the pioneers of modern management. His contribution to the practice of factory discipline had been the subject of a well-known article by McKendrick.[27] It was Wedgwood's involvement with the Trent and Mersey Canal (opened in 1777) which gave rise to the documents examined by Davies. As Ward had previously pointed out, his attitude to the company was not free from ambiguity. Although he had taken a prominent part in campaigning for a canal, from which he obviously stood to benefit as a shipper, Wedgwood did not appear as a shareholder until the early 1780s.[28]

The Trent and Mersey was the second largest canal to be built in England and Wales. Day-to-day management was normally in the hands of salaried officers rather than proprietors, but it seems that here the directors took a more active part than most. One distinctive early decision was to set up their own carrying firm, Hugh Henshall, which was operating seventy five barges by the early 1790s.[29]

At a time of disappointing results in the early 1780s, Wedgwood is thought to have drawn up a couple of memoranda setting out for the benefit of the directors his views on how the company ought to be conducted. The chief executive was to be known as the auditor, and was to exercise financial control, including the inspection of weekly returns from each wharf. There were also elements of public relations and marketing in his job description. The most striking suggestion was for the appointment of 'walking surveyors', who were to enjoy direct access to the directors and to act as their eyes and ears. Described by Davies as 'foremen, inspectors and trouble-shooters',[30] they had duties which left no doubt as to Wedgwood's lack of confidence in the efficiency and honesty of many employees. Another proposal was that the captain of each boat should keep a waybill - a kind of log-book which would be available for inspection by company officials. A specimen waybill reproduced in the article was probably fictional, designed to impress the directors with the potential advantages of the scheme.

These ideas were never implemented, and indeed the financial difficulties which prompted them were soon to disappear as the Trent and Mersey entered a period of higher profits and dividends. Davies speculated plausibly that Wedgwood's proposals would not have been successful in practice. It may be that he presumed excessive similarity between the problems of managing a factory and those of a transport undertaking. His thinking would, in any case, have had limited applicability to most inland navigation enterprises. The Trent and Mersey was atypical because of its size and the company's commitment to carrying as well as to canal operation. Parts of Wedgwood's memoranda seem to have been directed to the former activity. Most canal firms relied basically on collecting tolls from independent barge-owners, many of whom operated a single vessel. On the other hand, some integral aspects of canal management were not covered in the Wedgwood documents. These include the vital issues of pricing, which involved complicated and often secretive negotiations over tolls and charges with interested parties. As Turnbull showed in his monograph on Pickfords, these matters were frequently the most difficult and controversial tasks confronting canal companies and their customers.[31]

One problem which canal directors had to face at an earlier stage was how to find surveyors and engineers with the relatively novel skills necessary to establish the best route and to solve whatever problems of physical construction the route might pose. Those who could achieve this without serious cost overruns were especially invaluable. Hadfield, in 1953, approached these questions from the standpoint of the engineer rather than the employer. His study of the career of James Green was very much the tribute of a confirmed admirer, yet it was fair-minded enough to present sufficient material on which a less favourable assessment might be formed.[32] The setting for Green's endeavours was the hostile terrain of south-western England, which meant that even over short distances canals frequently required either unusually numerous locks, or some more adventurous method of traversing hilly ground. Hadfield stated at the outset that Green 'solved in terms of his medium' this problem, and that he developed his ideas, which were admittedly not new, 'more practically than his predecessors.'[33]

Green did a survey in 1817 for the Bude Canal, constructed between 1819 and 1825, and later claimed, though this is not verifiable, to have built it. Among the features for which he was presumably responsible were several inclined planes, mainly powered by waterwheels except for one which used an unreliable piece of machinery (sometimes replaced by a steam-engine) involving two large buckets in wells.[34] He also worked on plans for a ship canal between the English and Bristol Channels (never, of course, completed) and a technically orthodox extension to the Exeter Canal. Other credits included an inclined plane operated by water on the Torrington Canal.

Green's best known creations were the lifts on the Grand Western Canal. He had originally intended to rely entirely on inclined planes at the steepest points. These were more common devices, which had been installed in several locations in both Shropshire and the South-West. But Green changed his mind in favour of lifts. Hadfield wrote later that the set of eight designed for the Grand Western (and in use between 1838 and 1867) were the only examples of vertically - rising lifts counterbalanced by water which were successfully operated under working conditions.[35] A few previous experiments with lifts had been carried out, but may not even have been known to Green. His work on the same canal was, however, somewhat accident-prone. The biggest of the lifts partly collapsed because of subsidence, and one inclined plane simply did not work. Green had made an error regarded by Hadfield as 'inexplicable' in underestimating the amount of water needed to work the machinery.[36] A steam-engine was later installed. Not surprisingly the directors were unfavourably impressed, and dismissed Green in 1836, two years before the Grand Western opened. He did no more work on canals.

These incidents confirm that Green was an engineer with a predisposition to experimentation and high-risk methods. One suggestive detail in Hadfield's article concerns the report by Green for a Looe to Liskeard link in 1823, in which he recommended inclined planes to accomplish the ascent. The canal was actually built a few years later by a local engineer, who used orthodox locks with considerable success. Green's claim to fame rests more heavily on his patchy record as a technical innovator than on his contribution to the commercial prospects of the companies he served. Hadfield continued to place a high valuation on his achievements. In a later work he placed Green in a pantheon of 'the great British engineers' along with Brindley, the elder Rennie, Outram, the Whitworths, the Jessops, the Dadfords and Telford.[37] Hadfield's co-authored biography of William Jessop is likely to prove a better judged exercise in restoring the reputation of a relatively neglected engineer.[38] It should, however, be recognised that Hadfield was not alone in his enthusiasm for Green. Gladwin and White, in a technically well-informed survey, described him as 'a much maligned genius', explaining that 'the technology of his time was not equal to his plans', but accepting that 'in the main it cannot be said they were successful.'[39]

The article by Freer was one of a small number concerned with the living standards of canal-boat people.[40] It focused on the period between the end of the 'golden age' in 1840 (the onset of railway competition) and the Second World War. The economic background was one in which canals obviously suffered a fall in their share of the freight market, but nevertheless retained a virtually constant volume of traffic between the 1830s and 1914. Only after the First World War did a steep decline occur. Although canals were less profitable, increased productivity and reduced underemployment meant that wages were maintained more successfully than has often been assumed.

Freer contested the notion that canal work was a particularly impoverished occupation, and that railway competition had depressed earnings to such an extent that boatmen became obliged to resort to family labour, with wives and children living on board. She argued that boatmen's wages in the 1840s were considerably higher than those of agricultural labourers and roughly comparable with those of coal miners and cotton spinners. Even as late as the 1920s and 1930s, after some relative deterioration, men on the larger boats were paid significantly better than the average for the old staple industries. The percentage of boatmen living with their families on board varied between about 18 per cent and 26 per cent from 1850 to 1914. Furthermore this minority practice was established before the railway age, and was subject to regional variation, being to some degree a matter of local custom and individual tastes, rather than simply a dictate of economic necessity. From the 1870s, however, it came under increasing criticism, and by the turn of the century it was also true that standards of accommodation on board compared less favourably with those on shore. Freer conceded the important point that, despite the generally above-average wages earned by boatmen, a high social cost was paid by some of their families in terms of overcrowded and insanitary living conditions and educational deprivation.

The data offered in support of these arguments about real wages was admittedly fragmentary, and complexities in the payment system create formidable difficulties for any attempt to calculate average weekly earnings. It was accepted, for example, that 'almost no two trips ever attracted the same payment.'[41] As for health, Freer estimated that, apart from the indisputably high accident rate, boatmen enjoyed as high a standard on average as the rest of the population.

The main thrust of Freer's work was undoubtedly to revise upwards previous assumptions about the living standards of the canal labour force. These were probably not derived from specific historical studies of the subject, which were sparse in the extreme, but from the influence of the writings of George Smith of Coalville, the author of *Our Canal Population*, who had been instrumental in the passing of the Canal Boat Acts of 1877 and 1884.[42] Furthermore, awareness of the extent of the decline of canal traffic in the twentieth century probably encouraged an automatic backward projection of falling real wages into the nineteenth century. Almost the only modern academic investigation before Freer was Hanson's 1975 monograph. The findings on living standards were broadly compatible with Freer, although slightly less 'optimistic' in tone. Hanson had put more stress on long-term fluctuations in real wages, detecting improvement up to the 1870s, then deterioration from 1890. He also pointed to a decline in the numbers of children on board in the last decades before 1914.[43]

Sherwood's article was also concerned with living standards, which inevitably provided the background to the industrial dispute he investigated.[44] This is almost certainly the only recent academic study dealing with a single strike by canal workers in Britain. As the author acknowledged, this incident was of modest significance in the context of the national economy, given the high level and wide distribution of disputes in the years after the end of the First World War. It was, nevertheless, outstanding in the history of inland navigation,[45] and even more spectacular in its impact on the Midlands village of Braunston, which became the epicentre of the strike by virtue of its strategic position in the canal network.

The firm of Fellows, Morton and Clayton (FMC) was the main canal carrier operating between London and the West Midlands. It had expanded its activities in the late-nineteenth and early-twentieth centuries, and indeed continued to invest in its fleet even after the 1923 strike. But FMC found the going tough and competition keen in the early 1920s. Following a cut in freight charges to customers, it announced a substantial reduction in boatmen's wages in August 1923. This brought the company into collision with the recently formed Transport and General Worker's Union (TGWU). Contact probably occurred when union members met FMC employees during their visits to Brentford and Limehouse. At this stage the union's leadership was in expansive mood, hoping to establish its appeal to workers in all branches of transport. As part of its growth strategy the union executive was seeking national negotiations for the canal sector, rather than piecemeal dealings with the individual companies. Accordingly it drew up a programme of demands for canal workers. One key item was a proposal that carriers should agree to employ two adult males on each boat - implicitly an attack on the system of family labour wherever it still survived.

Sherwood's account of the strike contained some vivid detail, including a discussion of FMC's attempt to evict boatmen from strike-bound barges, and an outline of the emergency arrangements made for the schooling of a greatly enlarged number of children in Braunston during the strike. Judgements on the outcome are not entirely straightforward. Wages were cut, though not to the extent originally demanded by the employer (5 per cent, in two instalments, instead of 6.47 per cent). The strikers had remained solid for three months, the company had been obliged to negotiate with the union, and to accept arbitration. The TGWU appeared to be well on the way to establishing itself as a permanent presence on the canals. Although Sherwood recorded, no doubt accurately, disappointment at what was then dubbed an 'Irishman's rise' (a reduction in a pay cut), he did not take account of the context of falling prices in the early 1920s. Freer, six years later, did note that real wages had risen between 1921 and 1923.[46] It may be that what boatmen experienced in these years was, in the same terminology, an 'Irishman's reduction' - a cut in money wages which, because prices fell faster, left purchasing power higher, or at least as high as before. It is, however, clear that the TGWU was unable to achieve its ambitions of high membership and national negotiations, or to implement its long-term programme, including the elimination of family labour. The story of the 1923 strike suggests that these failures were attributable more to weakness of demand in both the canal sector and the national economy than to any absence of discontent among canal workers.[47]

De Jong's article is the only one devoted to a non-British subject.[48] It remains distinctive in that it offered a broad interpretation of the development of internal transport in a national economy, albeit a relatively small one. Inland navigation was not treated in isolation, but was related to other transport modes. The Netherlands had acquired an extensive canal network by the end of the seventeenth century, which was used almost exclusively by passenger traffic. The bigger rivers and the Zuider Zee were important for freight, with the latter acting as a large-scale traffic interchange, and carrying large quantities of peat.

In the nineteenth century there were three bursts of canal construction, in the 1820s, the 1850s and after 1865. The first wave mainly replicated disused natural waterways and was intended to restore the international trading position of the Dutch ports through improved links with the sea and with Germany, via the major rivers. The second wave attempted to connect hinterland areas with the main existing waterway network. All this canal-building did not open up any new urban or industrial areas, and did not lead to the creation of a national waterway network at any time before 1880. Nevertheless, at local level, differences in access to navigable waterways help to explain some variations in economic performance, such as the retardation of North Brabant and Tilburg.

Little was done to improve the navigability of the great rivers before 1850. But eventually the Rhine trade became an important economic influence, powered by German industrialisation and the development of a system of iron barges towed by steamships. Roads were relatively unimportant in economic terms, although political integration was significant as both motive and effect in their construction. By the 1860s and 1870s, the railways had captured most passenger traffic, while inter-regional freight continued to be dominated by inland navigation.

In the past it was commonly believed that the development of railways had been particularly slow in The Netherlands. One suggested explanation was the early strength and success of the canal system, and the inadequacy of the institutional machinery for transport innovation. This slowness was sometimes held responsible for overall industrial retardation. It seems on balance a more credible proposition that late industrialisation itself dictated the timing of railway-building, and that the pace of German economic development was highly relevant. Thus transport did not serve as a leading sector in The Netherlands.[49] A railway network was not laid down until industrial growth had shown both the advantages and the necessity of an integrated modern transport system. Furthermore the extension and integration of waterways after 1865 helped to confine the railways to a modest role.

De Jong did not make explicit international comparisons, but his survey inevitably served as a reminder of the extreme contrast between British and Dutch experience (Britain, for example, enjoyed no kind of canal revival after the early nineteenth century). It may perhaps stimulate further interpretive essays in the transport history of other economies, and research into the inter-relationship of road, river, canal and rail within different national contexts. Further comparative research is much needed, and is arguably essential to deepen our understanding of events in any single country.[50] It would certainly be satisfying if it succeeded in clarifying the reasons for the early decline of canals in Britain. Some traditional concerns, of course, remain valid: it would be desirable, sources permitting, to fill some of the gaps in our knowledge of the planning, construction and management of British canals, and of labour conditions before the late nineteenth century. But the limitations of the traditional approach have been apparent for some time, with its tendency to treat water transport in isolation. It is increasingly clear that transport projects of all kinds need to be studied in relation to supply factors, production and markets, and with due regard to the institutional setting. Within such a framework there is still scope for work focused on firm, industry or region. Most of the articles in this collection were written before (and indeed facilitated) the important recent contributions to the subject made by economic historians and economic geographers such as Pollard, Turnbull and Langton.[51] Their work has re-emphasised the importance of water transport in key regions during industrialisation. The work represented here is entirely compatible with such conceptions, and shows considerable sensitiveness to the local and regional context in tracing the impact of transport innovation on industries and communities.

Notes

1 C. W. Chalklin, 'Navigation schemes on the Upper Medway', *The Journal of Transport History*, vol. V (1961), pp. 105-15 (reprinted in this volume).
2 T. S. Willan, *River Navigation in England* (Manchester, 1936, 1964), pp. 16-18.
3 C. W. Chalklin, *Seventeenth Century Kent* (London, 1965), p. 166.
4 B. F. Duckham, 'Selby and the Aire and Calder Navigation 1744-1826', *The Journal of Transport History*, vol. VII (1965), pp. 87-95 (reprinted in this volume).
5 B. F. Duckham, *The Yorkshire Ouse* (Newton Abbot, 1967), pp. 72-7.
6 J. R. Harris, 'Liverpool canal controversies 1769-1772', *The Journal of Transport History*, vol. II (1956), pp. 158-74 (reprinted in this volume).
7 T. C. Barker and J. R. Harris, *A Merseyside Town in the Industrial Revolution: St Helens 1750-1900* (Liverpool, 1954).
8 T. C. Barker, 'The beginnings of the canal age in the British Isles', in L. S. Pressnell (ed.), *Studies in the Industrial Revolution* (London, 1960), pp. 1-22.
9 Harris, 'Liverpool Canal', p. 170.
10 J. Langton, *Geographical Change and Industrial Revolution* (Cambridge, 1979).
11 G. Firth, 'Bradford coal, Craven limestone and the origins of the Leeds and Liverpool Canal', *The Journal of Transport History*, 3rd series, vol. IV, no. 2 (1983), pp. 50-62 (reprinted in this volume).
12 Ibid., p. 53.
13 Langton, *Geographical Change*, p. 170.
14 J. R. Ward, *The Finance of Canal Building in Eighteenth Century England* (Oxford, 1974), p. 126.
15 Langton, *Geographical Change*, p. 230.
16 C. P. Griffin, 'Transport change and the development of the Leicestershire coalfield in the canal age: a reinterpretation', *The Journal of Transport History*, 2nd series, vol. IV, no. 4 (1978), pp. 227-38 (reprinted in this volume).
17 R. G. Wilson, 'Transport dues as indices of economic growth', *Economic History Review*, vol. XIX, no. 1 (1966), pp. 110-23.

18 C. Hadfield, *Canals of the East Midlands* (Newton Abbot, 1966), p. 86.
19 G. W. Crompton, 'Canals and the Industrial Revolution', *The Journal of Transport History*, vol. XIV, no. 2 (1993), pp. 106-7.
20 M. W. Flinn, *History of the British Coal Industry, II 1700 - 1830* (Oxford, 1984), p. 185.
21 H. Pollins, 'The Swansea Canal', *The Journal of Transport History*, vol. I, no. 3 (1954), pp. 135-54 (reprinted in this volume).
22 C. L. Baber, 'Canals and the economic development of South Wales', in C. L. Baber and L. J. Williams, *Modern South Wales: Essays in Economic History* (Cardiff, 1986), pp. 24-38.
23 S. Pollard, *Peaceful Conquest* (Oxford, 1981), pp. 3-19.
24 G. L. Turnbull, 'Canals, coal and regional growth during the Industrial Revolution', *Economic History Review*, vol. XL, no. 4 (1987), p. 544.
25 H. Morris and L. J. Williams, *The South Wales Coal Industry 1841-75* (Cardiff, 1958), pp. 93-4.
26 C. Davies, 'Josiah Wedgwood and canal management', *The Journal of Transport History*, 3rd series, vol. I, no. 1 (1980), pp. 49-57 (reprinted in this volume).
27 N. McKendrick, 'Josiah Wedgwood and factory discipline, *Historical Journal*, vol. IV, (1961).
28 Ward, *Finance*, pp. 78, 128, 134.
29 G. L. Turnbull, *Traffic and Transport: an Economic History of Pickfords* (London, 1979), p. 79.
30 Davies, 'Josiah Wedgwood', p. 52.
31 Turnbull, *Pickfords*, pp. 90-3.
32 C. Hadfield, 'James Green as canal engineer', *The Journal of Transport History*, vol. I, no. 1 (1953), pp. 44-56 (reprinted in this volume).
33 Ibid., p. 44.
34 Ibid., p. 47 for a detailed explanation.
35 C. Hadfield, 'Canals and inland waterways in the British Isles', in C. Singer, E. J. Holmyard, A. R. Hall and T. I. Williams, *A History of Technology IV* (Oxford, 1958), p. 569.
36 Hadfield, 'James Green', p. 55.
37 Hadfield, 'Canals', pp. 571-2.
38 C. Hadfield and A. W. Skempton, *William Jessop, Engineer* (Newton Abbot, 1979).
39 D. D. Gladwin and J. M. White, *English Canals II: Engineers and Engineering* (Lingfield, 1968), p. 48.
40 W. Freer, 'Standards of living among canal boat people 1840-1939', *The Journal of Transport History*, 3rd series, vol. XIII, no. 1 (1992), pp. 43-58 (reprinted in this volume).
41 Ibid., p. 45.
42 H. Hanson, *The Canal Boatmen 1760-1914* (Manchester, 1975), pp. 120-36.
43 Ibid., pp. 91-100.
44 K. B. Sherwood, 'The canal boatmen's strike of 1923', *The Journal of Transport History*, 3rd series, vol. VII, no. 2 (1982), pp. 61-79 (reprinted in this volume).

45 Hanson, *Canal Boatmen*, pp. 96-7 for examples of some previous strikes.
46 Freer, 'Standards of Living, p. 48.
47 H. A. Clegg, *A History of British Trade Unions since 1889 II 1911-33* (Oxford, 1985), chs 11 and 14.
48 H. J. de Jong, 'Dutch inland transport in the nineteenth century: a bibliographical review', *The Journal of Transport History*, 3rd series, vol. XIII, no. 1 (1992), pp. 1-22 (reprinted in this volume).
49 Crompton, 'Canals', pp. 106-7.
50 One important recent example is R. Szostak, *The Role of Transportation in the Industrial Revolution: a comparison of England and France* (Montreal and Kingston, 1991).
51 Crompton, 'Canals', p. 97-100.

1 *Navigation schemes on the Upper Medway, 1600–65*

C. W. CHALKLIN

During the seventeenth century many attempts were made to extend the navigation of English rivers. Among them were three projects to open up some fourteen to eighteen miles of the upper Medway. As far as Maidstone the river was already passable: now it was proposed to make it navigable up to Tonbridge, or even to Penshurst, the highest possible point. The practical results were almost negligible. Despite ambitious plans for building locks, widening the river, and constructing wharves and cranes, no more was done than the removal of some of the weirs. Yet the schemes are of considerable significance, for they reveal the clash of interests between the promoters and their opponents. On the one hand the ironfounders and timber-growers of the Weald of Kent were seeking a better means of carriage for their products to London, Chatham, and the north Kent ports than was offered by the roads, and the government a speedier delivery in wartime of the cannon and timber for its ships. On the other hand the local gentry and farmers along the banks between Maidstone and Yalding claimed that the river was as much their private property as the adjoining lands, and objected to the removal of the weirs and the making of towpaths.

The means proposed to make the river navigable are also interesting. At the beginning of the century the favourite instruments were the Commissions of Sewers, bodies of local landowners appointed by the Crown with responsibility for river drainage. But their powers did not properly extend to river navigation, and by the 1660s the general practice was to give statutory authority to a body of undertakers, who were to do the work at their own charge, usually in return for a monpoly of the carriage on the river. The Medway projects illustrate this change of method. The first plan, in 1600, was attempted by the local Commission of Sewers. At the time of the second scheme, in 1627, its inadequacy was realised; an undertaker made himself responsible for the work in return for a monopoly of the carriage, and an attempt was made to give him statutory authority. Finally the third plan began, and ended, with an Act of Parliament, which appointed ten local landowners as undertakers.

At this time the Weald of Kent and Sussex was one of the centres of the English iron industry, chiefly because of its large deposits of ironstone and the abundant supply of timber for charcoal. The making of iron began on a large scale in the 1540s, and by 1574 some hundred ironworks existed in the area. There were cast-iron products such as cannon, firebacks, and hollow-ware, and wrought iron for the use of smiths and cutlers. The principal markets were London and the north

Kent towns and dockyards. The Weald was also a main source of English timber, for medieval settlement had been scattered and much of the original forest remained. The wood was used both for fuel and general building purposes in London, and for shipbuilding in the dockyards.

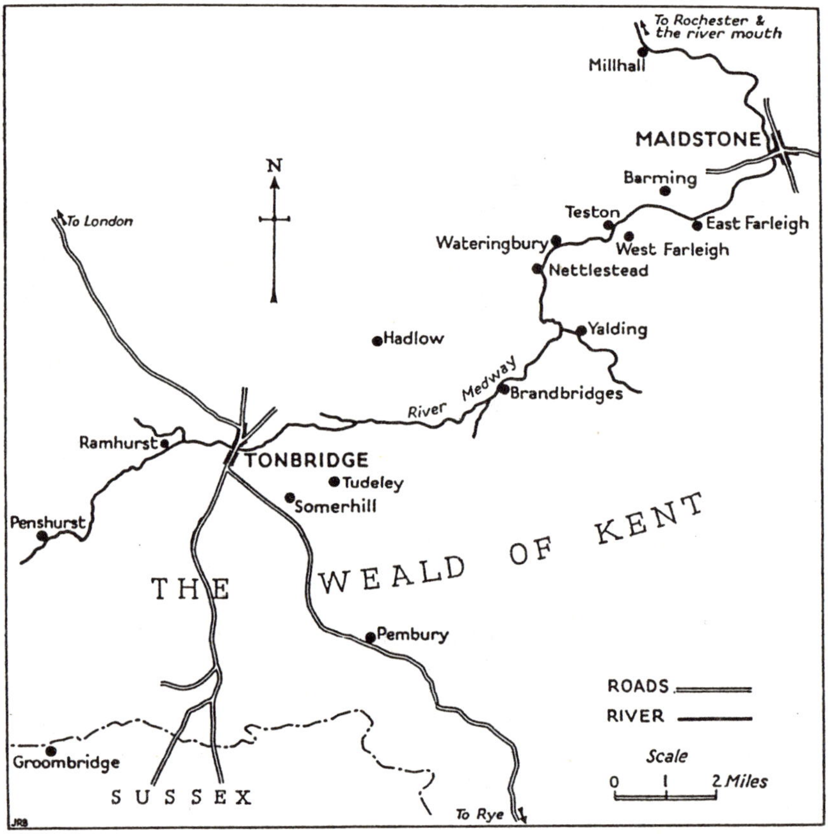

There were two routes for the carriage of these products to London and the north Kent ports. The first, to London, was the road which came ultimately from the Sussex port of Rye, and crossed the Medway at Tonbridge. The other route lay by minor roads to the River at Yalding or Maidstone, and thence by boat to Chatham, Woolwich, or London. The condition of the roads in the Weald was notorious. As they were unsurfaced tracks their quality depended on the soil and the weather. In a dry summer the clay might bake to a reasonably hard surface, but it became almost impassable in winter. Either in the summer of 1570 or 1571 a royal purveyor bought 93 trees out of Somerland wood, in Tonbridge parish; the tops and lops which he had intended to use for burning the Queen's bricks at Greenwich he sold locally, however, for with the approach of winter the carriage charges were more than could be allowed.[1] As the population and trade of the Weald increased in the late sixteenth century, the roads grew worse. The heavy loads of timber and iron were special causes of their deterioration. Two statutes, of 1584 and 1597, required ironmasters to contribute to the repair of the roads. Under the Act of 1597 they were required to pay the very large sum of 3s. for every three cartloads of coal or mineral, and for every ton of iron, carried one mile, between 12 October and 1 May, for the repair of the roads; and between 1 May and 12 October they were to provide a load of cinders and gravel for every 30 loads of coal or ore, or ten loads of iron being transported.[2]

With the growing difficulties of land carriage, transport by water became increasingly attractive for the ironmasters and landowners of the Weald. The growing use made of the river below Maidstone is shown by its development as a port during the reigns of Elizabeth I and James I. In 1565 a return to the Crown mentioned only four hoys belonging to the town; yet in 1619 its trade had become so considerable that the borough was granted in a charter the right to levy tolls for "wharfage, anchorage and groundage of all and singular ships and other vessels" coming to the town.[3] There was also an important wharf at Millhall, 3½ miles downstream. From at least 1580 small boats could intermittently travel upstream from Maidstone for six miles to Yalding. The aim of the supporters of navigation was first to make this section of the river passable. At times they also planned to make the Medway navigable eight miles further to Tonbridge, and even four miles beyond to the highest possible point, at Penshurst. Lying on the highway from Rye to London, as a small market town and the road centre of the western part of the Kentish Weald, Tonbridge was an obvious place for the embarkation and landing of goods.

It is possible that nothing would have been attempted in the early seventeenth century if there had not been in existence a body responsible for the proper flow of the river between Penshurst and the Thames. The Commission of Sewers was a body of local landowners appointed by the Crown. The Commission of 1627 consisted of 68 people: apart from two nominal members, the Lord Admiral and the Lord Chamberlain, the Mayors of Rochester and Maidstone, and one or two local noblemen like the Earls of Dorset and Leicester, they were country gentry. Their letters of appointment complained that "the walls, ditches, banks, gutters" of the river "be so disrupt, lacerate and broken . . . by means of setting up, erecting, and making of streams, mills, bridges, ponds, fishgarths, milldams" and spoke of the "great and inestimable damage for default of reparations"; the Commission was to survey and reform the abuses, and to levy a rate for the necessary works.[4] Its purpose was therefore the free passage of the river, not its navigation. At Yalding the Beult and Medway meet, having collected waters from a wide area of the

Weald plain and the hills beyond. The river down to Maidstone flows through a gap in the ragstone hills, and obstructions here caused serious flooding at Yalding and further upstream to Tonbridge. In 1600 it was said that "Tonbridge men do say, till the water be gone from Yalding, they are ever drowned, and as soon as that is gone, they are dry again".[5]

Between 1580 and 1597 three separate commissions had removed weirs, logs and other obstacles in the river above Maidstone to prevent floods, and some navigation had been possible.[6] Then the landowners along the banks between Yalding and Maidstone replaced the obstructions: they claimed that the river was part of their private property, and that the weirs protected their fish and were useful boundaries between farm and farm. Flooding at Yalding and upstream became serious. According to one eyewitness, whereas when the Medway was free of obstructions the area was never flooded for more than four or five days, the inundations now lasted three or four weeks or longer. There were reports that cottagers had to bring their swine into the ground floor of their dwellings and to live themselves, without fires, in the depth of winter, in the bedchambers. In 1600 the Commission of Sewers ordered the removal of the weirs. Almost at once there was opposition from the landowners beside the river below Yalding. They claimed that it was not the weirs but the ironworks in the Weald which were responsible for drowning the countryside; in time of rain the ironmasters emptied the huge ponds which they kept for the water power to work their bellows and their hammerbeams.

These landowners were, however, partly protesting against a scheme of several commissioners to make the river navigable at the same time. Its protagonists included Francis Fane of Badsell, Tudeley, and Thomas Waller of Groombridge Place, Speldhurst, both Wealden landowners probably concerned with the iron industry and the sale of timber, George Rivers of Hadlow, and Sir John Leveson of Halling. They claimed that the river had been navigable by prescription and belonged to the Crown as a highway. They stressed the improvement to the trade of the Weald that would result from the reopening of the Medway up to Yalding. Besides the salt, meat, butter, fish, and cheese that could be brought up the river, the cloth industry would benefit from the oil, woad, and madder that could come more cheaply than by land. They pointed out that the carriage of timber by land from the Weald was extraordinarily slow: according to one source timber would be felled in the spring and brought to Yalding during the summer. If water carriage was impossible the wood would lie there all the winter, unless the owner paid for its carriage by land to Millhall, the wharf near Maidstone, at twice the cost by water when boats were available. If it remained until the next summer it would still cost 4s. a ton to carry, and would not be available until the following season. Thus the shipowners had to have two years' provision of timber in store, for wood felled in 1598 could not be used until 1600.

Yet the landowners and farmers between Maidstone and Yalding were unwilling to surrender their interest in keeping that section of the river undisturbed. Quite rightly they claimed that the Medway had never been made navigable by Act of Parliament. This was the crux of their case. As this section of the river was non-tidal, and unaffected by statute, it was not owned by the Crown, and therefore not a common highway, but the private property of the adjoining landowners. Only an Act of Parliament, or royal Letters Patent, could take away their rights. The landowners also had several practical objections to navigation. The passage of boats up and down the river would damage the bridges; watermen respected no one's property, trampling down grass, stealing hens and breaking down fences. On 14 July 1600 the landowners' leader, Sir John Scott, of Scott's Hall, Nettlestead, asked the Lord Treasurer and the Lord Admiral to stay the proceedings of the commissioners. Realising there was a dispute among some of the leading landowners of the county, but without understanding the issues involved, they agreed. A meeting was appointed in London for 21 August, when both parties would be heard and a settlement reached.

The protagonists of navigation then tried to seek assistance at Court. A ready ally was soon found in the Lord Chamberlain, Hunsdon. In a letter of 20 July to Fane and the others he urged them to make the river navigable to Tonbridge and beyond. He did not attempt to conceal his personal interest: if the river was made navigable to Penshurst, Ramhurst mill above Tonbridge, the chief cause of the inundation of the surrounding land belonging to his ward, would be removed. The damming of the water at Ramhurst also often meant that his own mill at Tonbridge, where he was lord of the manor, was unworkable from lack of water. He no doubt presumed that since this mill lay on a small arm of the Medway it would itself be unaffected by the navigation. In his turn Fane urged the Lord Chamberlain to use his influence with the Lord Treasurer and the Lord Admiral. He stressed the improvement to his lordship's land at Tonbridge if floods were avoided, and the increased profits of his tenants if they could use the river for the export of their produce. Tonbridge would in fact become a "port" for the whole Weald: "your lordship's town of Tonbridge might be made a staple of all material for the Weald of Kent, for the enriching of that poor town and the general benefit of both shires". Hunsdon duly persuaded the Lord Admiral and the Lord Treasurer to give the scheme favourable consideration at the meeting on 21 August.

Yet instead of taking advantage of this powerful governmental support, Fane and his colleagues reached a compromise with Scott before the meeting, by which the navigation scheme was quietly shelved. On 19 August nineteen commissioners, including the supporters of navigation and Sir John Scott, viewed the river from Brandbridges down to East Farleigh bridge. At the end Scott agreed with the other commissioners that the obstacles such as weirs, trees, and logs should be cleared from the river bed; in turn the other party undertook not to meddle further with navigation, and to leave the matter to the law courts, Privy Council, or Admiralty to take such steps as they judged necessary. The next day Scott wrote to the Lord

Treasurer that he did not wish to attend the London meeting, as he felt that his own interests were now sufficiently protected. At this point the Fane correspondence about the Medway ends, and one assumes that the compromise was kept and the navigation scheme laid aside for a time.

The reason for this sudden partial surrender on the part of the commissioners is not given in the correspondence. Perhaps they realised that the limitation of their powers to the drainage of the river and the determined opposition of the local landowners was likely to outweigh the growing support for their scheme at Court. At least they were securing the removal of the causes of the floods of the previous winters.

No further attempts are known for 27 years. No doubt the long peace after 1604 lessened the demand for cannon and for timber for shipbuilding in the royal dockyards. The next project, in 1627, although once again revealing the deep cleavage of opinion between the landowners of the Weald and the owners and occupiers beside the river below Yalding, differed in several ways from the scheme of 1600. It was more carefully planned, and the struggle went on for several years. The landowners of the Weald, who were again among the chief protagonists of the scheme, realised that the Commission did not have the power to make the river navigable. While attempting to use it for the removal of weirs and other obstructions, which came within its responsibility for river drainage, they found a contractor to widen and deepen the river and prepare towpaths, in return for a monopoly of the carriage. He first tried unsuccessfully to secure statutory authority to override the rights of the local landowners; then he fell back on the power of the Privy Council. For instead of trying to act as arbiter, as in 1600, the government gave its active support to the plan throughout its course.

As in 1600, the scheme to make the river navigable in 1627 was accompanied by a demand for improved passage of its waters.[7] Early in the year there were once again complaints of flooding in the Yalding area owing to obstructions further down the river. On 14 April a Commission of Sewers was set up to remove all impediments in the river, making no mention of a navigation project. Yet its appointment coincided with governmental preparations for the French war, and there was immediate official pressure to make the Medway passable. At the first sessions of the Commission held at Tonbridge on 23 May to consider the obstructions in the river between Penshurst and Twyford bridge, at Yalding, letters were read from the King and the Duke of Buckingham. They had heard that the Medway up to Tonbridge could be made passable for boats to carry the products of the Weald at all times, whereas land transport was almost impossible for much of the year: "we are given to understand that there is a great quantity of timber, iron ordnance, and other commodities in those parts, which would come cheaper and more readily for the service of our Navy at Chatham all the year than now they are brought". The commissioners were therefore recommended to make the river navigable from Maidstone to Tonbridge for boats of good burthen, as it already was between Rochester and Maidstone.

There was strong support for the project from landowners in the Weald, such as the Earl of Clanricarde of Somerhill, Tonbridge, with an iron furnace and forge, Sir Peter Vanlore, owner of the Postern forge, Tonbridge, and Sir George Fane, of Badsell, Tudeley, with large timber interests. The Commission did not attempt itself to go further than order the removal of the obstructions on the river. Instead a contractor was found, Michael Cole of Westminster, gentleman, who came to an agreement with some of the owners and occupiers, presumably mostly from the Weald, at the meeting of the Commission at Tonbridge on 27 July. He recognised that their approval was necessary for making the river navigable and for the towing of boats from the banks. In return for a monopoly of the water carriage to London for 33 years, Cole was to make the river passable for boats of four tons between Maidstone and Penshurst, within three years. He was to buy four mills which obstructed passage, build sluices and locks where necessary, widen the river by digging away the banks, and make towpaths 4 ft. wide. The landowners were to be compensated for river widening or flooding. Within six months of the completion of the work Cole was to provide sufficient boats to carry everybody's goods from Penshurst to London once a week, and back in similar time. Goods were to be taken up at any suitable place on the river. A scale of maximum charges to and from London was prepared:

Loading or unloading place	*Maximum charge per ton for carriage to or from London*
Penshurst	10s.
Tonbridge	8s.
Brandbridges	5s.
Twyford Bridge	4s. 8d.

A clause said that "the lords and owners do declare that they will mediate with all other the said owners, whose names are not subscribed to this said agreement in every particular respect". No names are in fact given, for the manuscript contains only a copy of a draft of the articles, but one assumes that the landowners in the Weald above Yalding bound themselves to attempt to win the approval of the landowners and farmers further down the river. In fact they were unsuccessful. Opposition had appeared even before the agreement was signed. At a sessions at Maidstone on 20 June it was once again objected on behalf of the inhabitants below Yalding that the removal of the weirs was illegal and would mean the destruction of the fish, and that the navigation would mean damage to the adjoining land through widening of the river and towing on the banks. The swiftness and shallowness of the river between Maidstone and Yalding was a further difficulty, and if it was dug deeper the surrounding lands would be drowned. "The most part of the owners and occupiers of lands adjoining to the said river find themselves aggrieved with this work."

While Cole was to prepare the river for navigation by building the locks and widening the bed, the clearance of the weirs and other obstructions lay within the power of the Commission of Sewers as a drainage authority. At a sessions at Tonbridge on 27 September it was decreed that all impediments mentioned in surveys prepared by local juries on behalf of the Commission during the summer were to be removed before 20 May 1628 by the owners and occupiers along the banks. Yet when a sessions met at Boughton Monchelsea on 19 August only some of the weirs, logs, and trees had been cleared away. The opposition pretended that the remaining obstructions had not been noted by the juries. The Commissioners therefore themselves surveyed the river two days later and ordered their clearance.

Cole was also running into difficulties. He and his supporters among the Wealden landowners realised that if the river was made navigable by statute any rights that the landowners might claim to the privacy of their property along the banks would be destroyed. Thus in April 1628 a Bill was introduced into Parliament "for the making of the river of Medway navigable for barges, boats and lighters, from the town of Maidstone to the town of Penshurst, in the county of Kent". It was considered at a time when the quarrel between Crown and Parliament was at its height and monopolies were one of the chief questions at issue; thus the project may have suffered both because it had governmental backing and because it gave Cole exclusive rights. A committee voted the scheme a monopoly, and the session ended before the Bill could be recommitted. No further action would seem to have been taken.[8]

Despite this setback Cole began work on the river. In July he travelled up the river from Maidstone, ordering the occupiers to remove the bushes along the banks. At West Farleigh Augustine Skinner, a leading opponent of the scheme, forbade him to set foot on his land at his peril. On 2 January 1629, with two surveyors and hired workmen in a flat-bottomed boat and a large wherry brought from London, he took a day to get two miles above Maidstone. Some weirs belonging to George Carpenter and Roger Coveney, two other opponents, were destroyed. Next month he got some miles further. According to his enemies, his hired labourers cut down and mangled woods, and trod and pulled down fences. On the return journey the boat struck East Farleigh bridge, and weakened it by forcing out many of the stones.[9]

The support of the scheme by the government continued. On 12 June 1629 the Privy Council stayed the lawsuits begun by both sides, and sanctioned the navigation proceedings on the understanding that compensation was to be given to the owners and occupiers.[10] In the same month the Commission passed yet another decree, ordering the removal of obstructions between Maidstone and Yalding, on the lands of Sir Edward Scott of Nettlestead, Sir Thomas Style of Wateringbury, and Augustine Skinner, the leading opponents of the scheme, and other lesser offenders. On 27 July a sessions at Hadlow considered the nuisances on the other section of the river, up to Tonbridge. A petition by farmers and tradesmen of Tonbridge and Hadlow complained that floods still occurred, and another decree ordered the clearance of weirs, logs, and trees. Yet at a sessions at Maidstone on

25 May 1630 further floods and bad weather were given as an excuse for yet more delay.

It is however clear that since 1627 there had been some improvement, at least between Yalding and Maidstone. The opposition declared that many owners and occupiers had pulled up weirs and cut down woods through fear of the penalties in the decrees. In the summer of 1630 boats were travelling as far as Yalding. In a letter dated 20 July Sir George Fane of Badsell mentioned that the Treasurer of the Navy had granted a commission to a purveyor to take up 800 carts to carry timber purchased by the Crown at Badsell, presumably from himself, to the most convenient landing place on the Medway. Previously it would have gone to Millhall, but now it could be sent to Yalding. Cole presumably had barges on this section of the river; whereas the boats which had previously carried the timber belonged to the Crown, Fane now suggested that he should contract with Cole for its delivery.[11]

From this point our sources are fragmentary. Skinner and his confederates carried on the battle against the supporters of navigation. Probably in 1632 they petitioned th Council that in obedience to its instructions they had not impeded the work odfeavigation; yet they had received no compensation for the loss of their weirs, the nraining of their lands, and the cutting down of their woods.[12] In 1633 they were still trying to get legal confirmation of their claim that the river was a private stream and its soil was part of the lands of the adjoining owners. Two years later, following the removal of more obstacles, boats were passing five miles up the river from Maidstone with two-ton loads, and returning with 6½ tons. But again there had been hindrances from the occupiers: Robert Scoles, who held a small farm between Maidstone and East Farleigh bridge, had arrested two watermen who were bringing down ordnance, for towing on his land.[13]

Altogether the proceedings since 1627 had been very inconclusive. The river was never navigable above Yalding, and all that was achieved was the temporary reopening of the section from Yalding to Maidstone. Throughout the attempt the local landowners below Yalding had obstructed the scheme. They had been prepared to defy the authority of the Privy Council, and the project was therefore doomed to failure when Cole's Bill failed to become law.

Although there was little progress during the Civil War and Interregnum in the development of river navigation, in the 1660s there was extraordinary activity all over England. The promoters relied not on the authority of a Commission of Sewers, or Letters Patent, but of an Act of Parliament.[14] Among the statutes was an Act passed early in 1665 "for the making the River of Medway navigable in the Counties of Kent and Sussex". Once again the impetus came from the government and the inhabitants of the Weald, especially the ironmasters. During 1664 the country was drifting into war with Holland, and it seems likely that the demands of the dockyards were the immediate cause of the statute. The preamble stressed the advantage to the government of an easier means of transport of the Wealden timber and iron. The value of navigation to the trade of the Weald was also pointed out in much the same way as before, it being important "as well for carriage of iron ordnance, balls, timber, wood, corn and grain, hay, hops, wool, leather and all other provisions there had, growing and accruing from thence, as of coals, lime, stone, wares and all other necessaries and commodities thither, whereby commerce and trade there used will be much increased and the public weal advanced". Ten local landowners were appointed undertakers. They were headed by Lord Muskerry of Somerhill, Tonbridge, whose estate included over a thousand acres of coppice woodland and an iron furnace and forge. In return for a monopoly of the water carriage they were to make the river navigable at their own costs, by widening, deepening, and digging new channels, by the erection of locks, wharves, and cranes for loading and unloading, and even by the making of roads or tracks for the carriage of the iron and timber to the waterside. Nothing is said about the geographical extent of the scheme; probably the river between Maidstone and Yalding was again blocked, and it was intended to make the main river navigable up to Tonbridge.

In fact the scheme was never attempted; it was nearly 80 years before the project was tried again, and at last successfully. According to the supporters of the Bill of 1740 the plan of 1665 failed from lack of money.[16] The later scheme drew on the capital of 47 people, and perhaps the earlier project in 1665 would have been successful if it had had wider financial support.

Yet although this scheme was stillborn, it possessed one element from lack of which the earlier projects had foundered, that of statutory authority. The gentry and farmers along the banks below Yalding opposed the first two schemes in 1600 and 1627 because they involved the destruction of their weirs and damage to their lands. Since the river had never been made navigable by Act of Parliament, they had a strong case with their claim that it was not a highway, and that it was as much their private property as the adjoining lands. After 1627 their determined opposition, by starting lawsuits, ignoring from one pretext or another the orders of the Commissions of Sewers or of the Privy Council, or directly obstructing Cole and his workmen, proved too strong for Cole and the Wealden landowners, and the government which supported them.

NOTES

1. Public Record Office, Court of Requests (Req. 2) 286/14.
2. E. Straker, *Wealden Iron* (1931), 185, and 39 Eliz. I, c. 19 (1597–8).
3. J. M. Russell, *The History of Maidstone* (1881), 312.
4. Kent Archives Office, S/MN AZ1: "An account of the Proceedings of the Commissioners of Sewers towards making the River of Medway navigable between Maidstone and Penshurst in the County of Kent, with their decrees and orders for the removing the nuisances in and upon the said River and the branches thereof, 1627–30", 1–7.
5. B. M. Fane Correspondence: Add. MS. 34218, f. 50v.
6. The following account of the proceedings of 1600 is based on B. M. Add. MS. 34218, ff. 37–57.

There are brief accounts in W. T. Jackman, *The Development of Transportation in Modern England* (1916), I, 169–170, and T. S. Willan, *River Navigation in England, 1600–1750* (1936), 16–18. I would like to thank Dr. Willan for his many helpful comments on this paper.

7. Except where stated, the description of the project beginning in 1627 is based on Kent Archives Office S/MN AZ1. The reports of the juries which surveyed the river in the summer of 1627, and the decree of the Commission of 27 September, are also given in the Streatfeild MSS.: B.M. Add. MS. 34105, ff. 189–195. There is a short account in Jackman, I. 170–171.

8. *Journals of the House of Commons* I, 893, 895, 914.

9. B.M. Add. MS. 33923, ff. 42–44.

10. *Calendar of State Papers Domestic, 1631–3*, 480.

11. Public Record Office, S.P. 16/171/18. 12. *ibid.*, 16/229/45.

13. *Cal. S.P. Dom., 1635–6*, 64. 14. Willan, 28.

15. 16 and 17 Chas. II, private Act. 16. *Journals of the House of Commons*, XXIII, 443.

2 *Selby and the Aire and Calder Navigation, 1744–1826*

B. F. DUCKHAM

During its long and interesting history the Aire & Calder Navigation exercised a profound influence on several Yorkshire towns and cities: on the growth of Leeds and Wakefield in the eighteenth and early nineteenth centuries, on the rise of Goole from the 1820s, and on the prosperity of Selby between 1774 and 1826. It is the purpose of this article to examine the important relationship which existed between the navigation and Selby during the half century beginning 1774.

Before this date the Rivers Aire and Calder had been improved by a group of undertakers under an Act passed in 1699,[1] when the growing woollen towns of Leeds and Wakefield had been linked by waterway with the Ouse Navigation. Legally the undertakers had control over the two rivers from those towns to the confluence at Castleford, and from there to Weeland, near the point where the Aire falls into the river Ouse. They had no power to interfere with the estuary of the Aire below Weeland, and this tidal section soon became notorious for delays. It was the pressure of traders' complaints and memorials from towns in Yorkshire, Lancashire and Lincolnshire that precipitated the crisis of 1770–4.

Although the many mills throughout the Navigation's length were a constant source of annoyance to vessels, the chief grievance of carriers was the unimproved tideway of the Aire. Delays of several hours while stranded lighters waited for the next tide were commonplace; interruptions to navigation lasting a week or more were not unknown.[2] When John Smeaton reported on the waterway at the invitation of the proprietors in 1771 he found "the open river below Haddelsey [*sic*] lock, for nearly three-quarters of a mile [was] embarrassed with continual shoals".[3] Smeaton recommended several new cuts to eliminate the more difficult stretches of the Aire and Calder, but his most sweeping suggestion was for a short canal from Haddlesey to Gowdale Lodge Clough—a distance of just under four miles—to by-pass the worst portion of the tideway. This would have formed a loop on the south side of the Aire; subsequently William Jessop suggested a longer cut to the north as an alternative. This would have run from Haddlesey to Brier Lane End, near Newland, and had Smeaton's support.[4]

The undertakers attempted to obtain Parliament's permission to embark upon the suggested improvements, but met with failure in 1772 and the following year. Opposition from local landowners—some themselves proprietors—was intense,[5] while the position was complicated by the Leeds & Liverpool Canal Co. projecting a canal from Leeds direct to Selby. The route was actually surveyed by Brindley and Whitworth, and petitions in its favour were filed by many Yorkshire towns.[6] The Aire & Calder proprietors successfully contrived to resist the scheme, for its acceptance by Parliament would certainly have meant the end of the older navigation.[7] In 1774, therefore, the undertakers evolved a new plan, the chief part of which was to cut a canal from Haddlesey to Selby. This had the virtue of meeting the complaints of most of the landowners, and at the same time connecting the old navigation to the Ouse by an improved channel. In fact it promised more than the original designs, since a canal to Selby meant that the tidal Aire would be completely by-passed. A forceful document was produced extolling the merits of the idea and exposing the weaknesses—real and imagined—of a Leeds–Selby waterway.[8] To accompany these arguments with action the proprietors ended the lease on the Aire & Calder and took the navigation back into their own hands so that they could impress Parliament with their zeal.[9]

The royal assent was given to the Act[10] embodying the scheme for the Selby Canal in June 1774, despite vehement opposition from the corn and fulling-mill interest. Henry Baynes of Knostrop complained bitterly to Viscount Irwin while the Bill was before Parliament:

> "I am afraid your Lordship will think me impertinent in troubling your Lordship with so much of my scrawl; But I realy [*sic*] cannot help it, for since I have seen the Proprietors' new Bill I think they have the impudence of highwaymen; I find in the 3rd page at the Bottom, they want it hereafter to be left at their discretion whether or no they shall not have the power to come and plague us again. I think if the Parliament suffer such an Act to pass, they deserve to have another Guy Fox [*sic*] under their houses to blow them up. We choose Members for the Security of our properties and instead of answering that end, they act quiet [*sic*] the contrary and put them in the power of a parcel of Wolves."[11]

However, the Act did take the wishes of most landowners into account. One clause, for example, made special provision for the water level of the Selby Canal to be permanently two feet lower than the level of the Duke of Ancaster's land. Other landowners were reassured during the committee stage about the dangers of flooding.[12]

From the minute books of the Aire & Calder Navigation, and from certain other sources, it is possible to construct some account of the making of the Selby Canal and its impact on the town. William Jessop was appointed engineer, and it was his duty not merely to link Selby with the Aire, but to superintend the several new cuts which were to be made on the two rivers themselves. Upon appointment he came before the navigation committee and swore what almost amounted to an oath: "I mean to make this Business", he told the proprietors, "the principal object of my attention and to give it every Degree of Attendance that may be necessary to compleat [*sic*] it to the very best manner I am able for the Advantage of the

Proprietors and to my own Credit."[13] He was paid a salary of £250 per annum while in the navigation's service, and it is clear from the minutes that he was expected to treat with local landowners and even negotiate compensation.[14]

The canal was dug to give a mean draught of 3½ ft., which was the depth of the old navigation before the improvements were implemented under the Act of 1774. It proved to be a serious miscalculation, for the shallowness of the canal was one of the reasons for the Aire & Calder Navigation's subsequent decision in 1820 to build a completely new waterway from the Aire to the Ouse.[15] Although one may blame the proprietors for a lack of foresight it is doubtful whether anyone in 1774 could have accurately predicted the enormous increase in the volume of water-borne trade during the next fifty years. Even Jessop was convinced of the canal's adequacy to meet the demands made upon it.[16] Certain arrangements made for drainage—such as culverts under the canal—made it difficult later for the undertakers to deepen the channel. A report made in 1797, when the waterway had been in use for only two decades, commented: "The Selby Canal is the only deficient Part of the Navigation; it was originally made too Shallow not having more than three feet six Inches Water, to remedy which, double Heights of Deals are obliged to be kept up at Haddlesey Dam."[17] The expedient of increasing the quantity of water in the channel by these artificial means was probably unpopular locally (because of the fear of flooding) and may well have had a detrimental effect on the canal banks. The cost of the cut when completed appears to have been about £20,000.[18]

The formal opening of the canal in 1778 was attended by some ceremony. "On Wednesday [29 April] . . . the canal, lately made at a very great expence [*sic*] by the undertakers of the Air [*sic*] and Calder Navigation, from Haddlesey to Selby was opened, and several vessels passed through to Selby, on which occasion a large concourse of people assembled, and the same was conducted in a very great order with firing of cannon, music and ringing of bells to the great satisfaction of the gentlemen and others at Selby and that neighbourhood; by which canal a much easier communication is opened between the towns and collieries adjoining the rivers Air and Calder, the City of York, and all other towns bordering on the rivers Ouze, Derwent, Etc., and will save upwards of twenty miles distance navigating."[19] In hard commercial facts there is little doubt that the carriage of most goods to and from the West Riding became cheaper for Selby.[20]

The effects of the opening of the canal were far-reaching for both Selby and the navigation as a whole. The town of Selby, it must be remembered, was already no stranger to water transport. Standing as it did on the navigable river Ouse at a point where a regular ferry operated, it had long been a port of call for small coasters and river craft. The river, being tidal to a point well above the town, was fairly easily navigated as far as Selby. It is true that some of the tide-flow had been absorbed by Vermuyden's Dutch River which drained the levels between Doncaster and Goole, but this had been injurious to the trade of York rather than that of Selby.[21] The conservation of the Ouse was admittedly in the hands of York Corporation—it had been so for three hundred years—but at least no tolls were payable below the mouth of the Wharfe, about six miles above Selby.[22] In practice this meant that Selby traders enjoyed a free river below the town, though they might not always be entirely satisfied with its upkeep. The coming of the canal, then, linked Selby's coastwise trade to the rich hinterland of the West Riding. The town's position was consequently established as that of the principal port on the river Ouse.

In July 1778 the Aire & Calder Navigation approached Lady Stourton (who owned much of the riverside at Selby) to ascertain whether she would sell "10 or 12 Acres of Land adjoining the Banks of the Ouse at Selby for Building Warehouses . . . in order to establish a large and extensive Trade there".[23] An agreement was soon concluded between the undertakers and Lady Stourton whereby the navigation was allowed to acquire some eight acres on the right bank of the Ouse just below the point where the canal locked into the river. The land was not in fact purchased, but obtained on a 99 years' lease.[24] Six years later, after most of the warehouses had been built, a cut or channel running parallel to the river was made from near the canal lock.[25] Associated with this were a dry dock, counting house, crane, warehouses, rigging house, tarring house, sailmaker's shop "and a Place for depositing Old Ropes".[26] In short, the navigation had provided itself with an important terminal depôt equipped not only for the handling of goods, but also for the repair of barges and ships.

There are many later references in the minute books to the extension of these facilities and to the growth of a lively transhipment trade in Selby. By 1788 we hear of the lock keeper being provided with an assistant "in consequence of the great Increase of Trade . . . at Selby".[27] In 1790 an extra clerk was needed there by the navigation,[28] and further examples could be given. The nature of the trade resulted chiefly from the disparity of the two waterways—the Ouse and the canal. Use of the canal was restricted by its size to barges and keels of up to 60 tons or so;[29] the Ouse, on the other hand, was capable of accommodating much larger vessels even above Selby. A variety of boats plied to and from the town including brigs, schooners, sloops, and Humber keels. The largest vessels were normally around 200 tons, and occasionally slightly larger.[30] It was thus at Selby that Yorkshire's share in the coasting trade began and ended, for above the town navigation soon became difficult for most sea-going ships.[31] This is not to deny the existence of a certain amount of through trade by lighters between Hull and Leeds or Wakefield. Many goods were, however, transhipped from coaster to barge and *vice versa* at the new wharves and docks of Selby.

By the early nineteenth century upwards of 800 coasters were being cleared at Selby each year.[32] A contemporary estimated that 369,780 tons of goods were actually handled in the port during 1800.[33] Most vessels brought or carried away goods which were neither locally produced nor meant for local consumption,[34] but this is not the whole picture. Cheaper transport gave a boost to the town's own commerce. Flax dressing, linen manufacture, rope-making, sail-making, corn milling, and even ironfounding prospered.[35] Shipbuilding, already an ancient industry in Selby,[36] enjoyed a boom period, and several local families—Weddall, Staniland, Audus—considerably increased their fortunes from the maritime trade and its needs.[37] The town grew physically larger, and the increase in population from 1801 (2,861) to 1821 (4,097) was proportionately larger than for any other twenty-year period during the nineteenth century. The chief local export was stone, sent coastwise down the Ouse, and a directory for 1834 mentions the names of no fewer than five agents for the trade.[38]

Shortly after the conclusion of the Napoleonic Wars steam made its appearance on both the Ouse and the Aire & Calder system. Steam packets were extensively used for passenger transport before the railway age, and Selby's position at the junction of two great waterways ensured for it an important rôle in this traffic. The toll bridge had, moreover, replaced the old ferry in 1792, and the town had developed some coaching trade. It is probably an exaggeration to say that "previous to the establishment of these vessels [i.e. steam packets], scarcely a stage coach was ever known to enter the town",[39] but it is certain that the two forms of transport stimulated each other. At Selby the traveller from the West Riding towns could change from coach to steamer for Hull. Indeed it is interesting to note that the original intention of the later Leeds & Selby Railway was "for making a Railway from the town of Leeds *to the river Ouse*, within the Parish of Selby".[40] Thus the railway at first replaced only the coach. Later the packets themselves were driven out of existence except for occasional pleasure cruising.[41]

About 1825 Selby's river-borne trade was at its height. Steamers plied daily on the tide to Hull and back, several fly-boats operated each day between Selby and Leeds, and there was regular water connection with York, Gainsborough, King's Lynn, Yarmouth, and Wisbech, and with London.[42] A branch customs house had been established, and vessels engaged in the coasting and near continental trade no longer needed to be detained and cleared at Hull. Although shipping was occasionally subject to delays on the Ouse, the carriage of goods was generally efficient and brisk, the journey to Hull by steam averaging some five hours.[43] This very prosperity, however, had emphasised the inadequacy of the Selby Canal. Despite the increased trade a cloud had hung over the town's commercial future since 1820 when the undertakers of the Aire & Calder had obtained an Act to cut a wider and deeper canal from Knottingley to Goole.[44]

In the meantime the Aire & Calder Navigation had enjoyed half a century's expansion helped by the Selby Canal. It was only when it became apparent that a much more reliable "main line" was required that the plan emerged to by-pass the Aire *in toto* below Knottingley, thus providing an entirely new route to the Ouse and Humber.[45] Figures for the period under review bear witness that the partnership between the navigation and Selby had not been solely in the latter's favour:[46]

Aire & Calder Navigation

Year	*Turnover*	*Total dividends paid*
1780	£41,432	£12,000
1785	£53,673	£17,000
1790	£64,915	£27,000
1795	£69,642	£32,000
1800	£126,292	£34,000
1805	£147,135	£46,000
1810	£129,758	£48,000
1815	£150,747	£43,200*

* After deduction of tax

Personal fortunes had been amassed by many of the landed gentry who were proprietors of the navigation. For some the dividends were merely a useful supplement to the income from their agricultural estates; for others they formed part of a complex of revenues which included mining rents and coal royalties.[47] In 1826 the two largest proprietors, Richard Fountayne Wilson and Sir William Milner, earned £9,322 and £7,843 respectively from their navigation holdings.[48]

That the corporate strength of the undertakers had grown commensurately with the increase in the water-borne trade is well attested by Thomas Hill of Leeds. Writing of opposition to part of the Act of 1820, he observes:[49]

> Those gentlemen who corresponded with me during the progress of the late Bill, in 1820, in Parliament, and took an active part to obtain the rejection of the arbitrary unwarrantable clauses, and the introduction of the protecting ones, can well . . . [confirm] the anxiety, application, and almost insurmountable objects we had to contend with to make any head at all against the overweening influence of such an opulent body as the Proprietors of the Aire & Calder, who had half a dozen Members of Parliament amongst themselves, and a great influence with the County Members.

The Knottingley–Goole Canal, originally projected and designed by Rennie, but built by George Leather, was completed in July 1826. As Hill notes, it had been opposed in part by some members of the landed interest—notably by the Hon. Edward Petre, then lord of the manor of Selby—but in the long run it was welcomed by almost all South and West Yorkshire traders and coalowners.[50] Its effect on the town of Selby was naturally serious. Whilst Goole, previously "an obscure hamlet",[51] grew into a large river port visited by some 2,000 ships in 1850, Selby lost its position in the coasting trade.[52] Local traffic on the Ouse indeed remained and the town was still a principal port of call for steam packets. The transhipment trade, however, was soon a thing of the past. The new canal from Knottingley to Goole had closed one chapter and opened another for both Selby and the Aire & Calder Navigation alike.[53]

NOTES

1. 10 and 11 Wm. III, c. 25. For the earlier history of the Aire & Calder Navigation, see T. S. Willan, "Yorkshire River Navigation 1600–1750", *Geography*, 1937; G. Ramsden, "Two Notes on the History of the Aire & Calder Navigation", *Thoresby Society Miscellany*, vol. XII, part 2 (1954); B. F. Duckham, *Navigable Rivers of Yorkshire—their History and Traditions* (1964), 9, 15, 19–21. For the passing of the original Act see R. C. Unwin, "The Aire and Calder Navigation, Part I: the Beginning of the Navigation", *Bradford Antiquary*, XLII (1964), 53–85.

2. *The* REPORT *of John Smeaton, engineer, upon the means of improving the navigation of the Rivers Aire & Calder, from the free and open tidesway to the towns of Leeds and Wakefield respectively*, 28 Dec. 1771: *Smeaton's Reports*, II, *passim*.

3. *Ibid.*, II, 137.

4. William Jessop, sometime pupil of Smeaton, had been called in by the undertakers in 1772. His decision was confirmed by Smeaton in December of that year: *Reports*, II, 151–3.

5. John Stagg of Selby, for example, wrote to the Duke of Ancaster's secretary on 8 Jan. 1773: "I need not add, no Time shd. be lost. . . . The proprietors at the meeting were unanimous to join Lord Irwin, other Nobles and Gentlemen to prevent any alterations of the Rivers Air [*sic*] & Calder from Leeds to Haddlesey and so forward to Newland; And signed a proper Instrument to contribute toward the Expence [*sic*] to prevent the Same". *Temple Newsam MSS.*, TN/LA5/1/24. Previously Sir William Milner, a leading member of the undertakers, had tried to prepare Lord Irwin for the decision of the navigation. Milner to Viscount Irwin, 23 Dec. 1771, TN/LA5/1/4. It may be noted that Lord Irwin owned considerable property adjacent to the river Aire.

6. *Journals of the House of Commons*, XXXIV, 115.

7. J. Priestley, *Historical Account of the Navigable Rivers, Canals and Railways Throughout Britain* (1831), 11. See also W. W. Morrell, *History and Antiquities of Selby* (1867), 174–5.

8. *The* CASE *of the Undertakers of the Old Navigation of the Rivers* AIRE *and* CALDER, 21 Feb. 1774. The chief advantage of the Haddlesey–Selby canal over its rival was its shortness (under six miles) and therefore its cheapness.

9. It had previously been the practice of the undertakers to lease the waterway's tolls for an annual rent. The original rent obtained (1704–5) was £800; in 1774 the lessee had been paying £8,500, Ramsden, *loc. cit.*, 384. The first year under the proprietors themselves produced a turnover of £40,232; by 1778 £12,000 p.a. was being paid in dividends alone: *Treasurer's Book*, 1775–1816, British Transport Historical Records (York), ACN 4/36.

10. 14 Geo. III, c. 96. "An Act to Amend an Act passed in the Tenth and Eleventh Years of the Reign of King *William* the Third, intituled '*An Act for the making and keeping navigable the Rivers of Aire and Calder, in the County of York*', and for improving the Navigation of the said River *Aire*, from *Weeland* to the River *Ouze*, at the *Old Brick Garth* at *Ouze Gate End*, within the Township of *Selby*, in the said County, and for other Purposes." Selby interests had opposed the Act of 1699, but were strongly in favour of this measure.

11. Baynes to Viscount Irwin, 27 Mar. 1774, *Temple Newsam MSS.*, TN/IA5/1/31.

12. *Journals of the House of Commons*, XXXIV, 478. Jessop told the committee that the canal would be provided with a double pair of flood gates at each end.

13. Minutes, 4 July 1774, ACN 1/5.

14. The salary is identical with that paid by the Calder & Hebble Navigation to John Smeaton when that waterway was constructed, 1758–65: Duckham, "John Smeaton and the Calder & Hebble Navigation", *Journal of the Railway & Canal Historical Soc.*, X (1964), 19–25.

15. See p. 27.

16. *Journals of the House of Commons*, XXXIV, 478.

17. *Report to a General Meeting of the Undertakers*, 17 Aug. 1797, ACN 1/6. The report also refers to the unsatisfactory state of the canal banks.

18. Preface to the *Treasurer's Book*, 1775–1816, ACN 4/36.

19. *York Chronicle & General Advertiser*, 8 May 1778.

20. 14 Geo. III, c. 96. Under the earlier Act tolls could be as high as 10s. per ton for goods carried from the Ouse to Leeds or Wakefield. There was also a considerable saving in time, vessels being able to accomplish the journey to Leeds in about twelve hours instead of needing twice that time. See *Journals of the House of Commons*, XXXIV, 479. The tolls under the new Act may be summarised thus:

To or from Selby *per ton per mile*	
Dung, stable manure, coal, cinders, slack, charcoal	½d.
Lime	¾d.
From Selby to Leeds or Wakefield *in toto*	
8 bushels (Winchester measure) of wheat, rye, oats, barley & other grain, beans, malt, rape seed, mustard seed, line seed	6d.
32 pecks of apples, pears, onions, potatoes	9d.
Bags of wool (not exceeding 312 lbs.)	10½d.
Chalk, fuller's earth, pig iron, kelp, flints, pipe clay, sand (except that obtained from the rivers Aire & Calder), stone, bricks, whiting, rags, old ropes, lead, plaster, alum, slate, old iron, tiles, straw, hay, British timber *per ton*	3s. 0d.
Fir timber, deals, battens, pipe staves, foreign oak, mahogany, beech logs *per ton*	3s. 6d.
Flour, copperas, woad, tallow, ashes *per ton*	4s. 0d.
Bad butter (grease) *per ton*	4s. 3d.
Powder-sugar, currants, prunes, brass, copper *per ton*	4s. 8d.
Treacle *per ton*	4s. 9d.
Soap *per ton*	5s. 4d.
Bar iron *per ton*	5s. 6d.
Cheese, madder *per ton*	6s. 0d.
Cloth, bales & all other goods *per ton*	7s. 0d.

21. F. Drake, *Eboracum: or the History and Antiquities of the City of York* (1736), 231. The tide-flow of the Ouse was in 1778 (as today) arrested at Naburn dam several miles above Selby. The dam had been built by York Corporation in 1757 to give a reliable depth of water between there and the city. For the Ouse Navigation see Duckham, *Navigable Rivers, passim*, and "The Navigation of the Yorkshire Ouse during the Nineteenth Century", *Journal of Transport History*, VI (1964), 182–8; also *V. C. H. Yorks.: City of York*, 472–5.

22. An Act of 1732 (Geo. II, c. 15) allowed the Trustees of the Ouse Navigation—the Lord Mayor of York, Recorder, Aldermen, Sheriffs, two members of the City Council, six of the city's Privy Council and twelve of the Common Council—to levy certain tolls only *above* the mouth of the river Wharfe.

23. Minutes, 6 July 1778. ACN 1/5. **24.** *Ibid.*, 10 Aug. 1778.

25. *Ibid.*, 5 Aug. and 1 Dec. 1784. The estimate for the cut was £322 6s. 6d.; that for the dry dock was £154 11s.

26. *Ibid.*, 1 July 1782. **27.** *Ibid.*, 7 Apr. 1788.

28. Minutes, 22 Mar. 1790, ACN 1/6. **29.** *York Courant*, 30 May 1820.

30. H. Schroeder, *Annals of Yorkshire* (1852), 149.

31. Duckham, *Navigable Rivers*, 25. It was the hope of attracting some coastal trade that persuaded York Corporation to build a larger lock at Naburn, 1886–8.

32. E. Baines, *History, Directory & Gazetteer of the County of York* (1822), I, 273.

33. J. Mountain, *Selby, Ancient and Modern* (1800), 159.

34. *The Case of the Petitioners for Erecting a Bridge over the River Ouse at Selby* (Jan. 1791) mentions the growth of the transhipment trade as one of the chief arguments for replacing the ferry: ". . . the manufactured goods of that Country [*sc.* the West Riding, Lancashire and Westmorland] which are sent all over the Continent of Europe, as well as for London and other places Coastwise, pass down that way [i.e. through Selby], and are there re-shipped, or put on board larger vessels to be conveyed to Hull and other places. In like manner all raw materials, as Wool, Cotton, Linen Yarn, Dying-Woods [*sic*], and a variety of other articles for the use of the Woollen and Cotton, Etc., Manufactories; as well as Groceries, Fir Timber and Deals, Etc. for the use of a great Part of the West-Riding of Yorkshire and a part of Lancashire, also pass that way upwards, and are there removed into smaller vessels to be carried up the different Navigations, connected with the River Aire".

35. Baines, *Directory*, 274–7; Morrell, *op. cit.*, 174–5.

36. W. Wheater, *History of the Parishes of Sherburn and Cawood*, 2nd ed. (1882), 210, mentions a ship, *Catherine of Selby*, which helped to carry Henry V's expedition to France in 1417.

37. Morrell, *op. cit.*, 174–5. Vessels up to 800 tons could be built in the Selby shipyards.

38. Pigot & Co., *National Commercial Directory* (1834), 914–5. 39. *Ibid.*, 914.

40. Leeds & Selby Railway Act, 1830 [italics mine]. The railway company subsequently built three jetties and a coal spout on the Ouse at Selby. The first shipment of rail-delivered coal was carried from Selby to Rochester by the schooner *Audus* in February 1835. See W. W. Tomlinson, *The North Eastern Railway—its Rise and Development* [1914], 254, 257, 260.

41. Duckham, "The Navigation of the Yorkshire Ouse", *Journ. Transport Hist.*, VI (1964), 182–8.

42. Baines, *Directory*, 278.

43. Sailing vessels were, of course, slower. Tomlinson, *op. cit.*, 257, notes that passengers on a steamer which grounded were expected to help the captain re-float her.

44. 1 Geo. IV, c. 39. As in 1774 the undertakers were moved to take action partly through fear. A Don–Aire canal had, for instance, been promoted by South Yorkshire land and coalowners. See Priestley, *op. cit.*

45. An additional Act of 1828 (9 Geo. IV, c. 98) enabled the navigation to effect further improvements.

46. *Treasurer's Book*, 1775–1816, ACN 4/36: Figures to nearest £.

47. This latter question is dealt with by J. T. Ward in several articles. See especially: "West Riding Landowners and Mining in the Nineteenth Century", *Yorks. Bulletin of Econ. and Social Research*, vol. XV, 1 (May 1963); "The Squire as Businessman: William Aldam of Frickley Hall (1813–1890)", *Trans. Hunter Archaeological Society*, VIII, 4 (1962); "West Riding Landowners and the Railways", *Journ. Transport Hist.*, IV (1960) 242–51.

48. Morrell, *op. cit.*, 174. The total dividend paid in 1826 amounted to £70,000.

49. T. Hill, *The Old and New Rates of Tonage* [*sic*] *. . . upon the Aire & Calder Navigations from Selby, Weeland and Goole to Leeds and Wakefield* (1826), 4.

50. *York Courant*, 10 and 24 Jan., 30 May 1820; 25 July 1826.

51. The phrase is borrowed from Priestley, *op. cit.*

52. Selby's coastwise trade did not, of course, entirely end; some remains to this day.

53. The writer wishes to acknowledge with gratitude his debt to the British Railways Board for their kindness in making available certain historical documents and allowing him to quote from them. A special word of thanks is also due to Mr. G. Thompson, City Librarian, Leeds, for permission to make use of the Temple Newsam muniments in his care.

3 *Liverpool canal controversies, 1769–72*

J. R. HARRIS

A CONTEMPORARY of the Canal Age wrote that "no towns have derived greater advantage from canals perhaps than Manchester and Liverpool."[1] Certainly the function of Liverpool as a canal terminal needs no labouring. What perhaps is not so clearly realized is the speed with which the port was connected to the national canal system, and the degree to which Liverpool's canal facilities were the creation of the first period of canal investment which ended in the boom of 1767–72.[2] This article seeks to do three things: to amplify the early history of the Merseyside canal system, and thus continue the narrative so far pieced together by Dr T.C. Barker and the writer;[3] to outline the various competing schemes dangled before the public between 1769 and 1772, including the almost forgotten project of the Liverpool Canal; and, finally, to emphasize once again the preoccupation of the citizens of Liverpool, from the highest rank to the lowest, with the question of coal supply.

Other more general reasons, such as the favourable nature of the flat Lancashire and Cheshire plains for canal building, and the previous experience won with the Mersey-Irwell, Douglas, and Weaver river navigations, must not be forgotten, but contemporary documents would show that the encouragement of canal building round Liverpool came principally from the great shortage of coal there, and from the welcome and patronage extended to such schemes by the Common Council of Liverpool.

The Sankey Navigation had been welcomed by the Corporation, and largely financed by its members. While the scheme enlisted public support by promising the town an improved fuel supply, the main intention of the principal promoter, John Ashton, was obviously to supply his Dungeon saltworks with coal. Another "undertaker" was John Blackburne, owner of the Liverpool salthouse; and the canal, by opening into the Mersey opposite the mouth of the Weaver, was clearly designed to supply coal to the Cheshire salt boilers also.[4] John Holt is explicit on this point, assigning "the want of coal to the saltworks", together with the badness of the roads serving them, as the "great motives" for the building of this, the first canal of the Industrial Revolution.[5]

To the salt boilers and to the Liverpool industrial consumers,[6] who would be prepared to hold considerable stocks of coal, the Sankey was no doubt a reasonably efficient communication; but the householders of Liverpool, who had looked to it for a supply of domestic fuel, did not find it so satisfactory. It is true that it proved more efficient than the former method of carting from Prescot and St Helens, and that its first effect was to put many of the carters out of business.[7] But the price of its coal rose from 7s. at the opening of the canal in 1757 to 8s. 6d. four years later,[8] the quality of the better

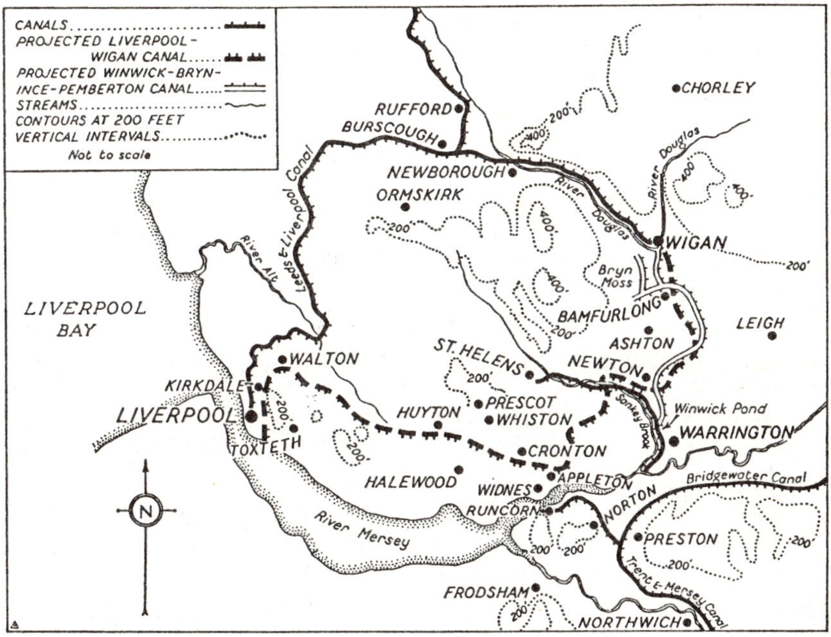

Principal Merseyside Canal Schemes, 1769–1772

grades of coal seems to have fallen off,[9] there were complaints of bad measure at the proprietors' coal yards in Liverpool, and the dealers seem to have been reluctant to sell in the small quantities which the ordinary householders desired[10]—a point on which the Duke of Bridgewater deferred to his poorer customers in Manchester and (later) at Liverpool.[11] The power of the great woman coal owner, Sarah Clayton, and her nephew, Jonathan Case, in the St Helens district, from which the majority of the coal came, seems to have aggravated the Sankey monopoly.[12] The canal, moreover, had many technical defects, the greatest being its joining the Mersey in the tideway, leaving the flats[13] to deal with the problems of wind, shallows, and tide, and seriously delaying onward transit to Liverpool. The proprietors, too, did not keep the canal in the best repair, and were slow in connecting it to the new mining district of Ravenhead outside St Helens, which was not reached till 1773.[14] Some measures were indeed taken to placate the Liverpool public,[15] but nevertheless great annoyance was felt, and vitriolic correspondence appeared in the Liverpool newspapers.

The Sankey canal scheme did not end the interest of either the Corporation or the public of Liverpool in transport improvement; if anything its very deficiencies spurred them to greater efforts. In 1764 the Common Council sent for Brindley to recommend a method of cleaning out their docks,[16] and in 1765 (once satisfied that it was for "the encouragement and promotion of trade") they contributed handsomely to the preliminary costs of the Trent and Mersey scheme, wrote to the M.P.s of the neighbouring counties in its favour, and assured the undertakers that they would give them "all the interest and assistance they can".[17] In the following year Liverpool was offered a scheme so far in advance of its time that it has not yet been carried out—that of a canal and road bridge over the Mersey at Runcorn. An immediate opposition from both Warrington and Liverpool, attacking it because of its possible interference with the navigation of the Mersey, suppressed the plan for the time being.[18] These various endeavours for transport improvement show that the canal fever of the late sixties and early seventies in Liverpool was but the climax of a long period of intense interest in improving the local communications system, and that the Common Council was favourable not simply to local schemes, but also to those of wider application, and appreciated to the full the possibility of extending the economic hinterland of Liverpool by inland navigations.

The Leeds & Liverpool scheme dominates the next phase in Liverpool canal history, which is now concerned with the controversies it excited and the rival schemes it inspired. A scheme of this nature was first mentioned in 1764, and by 1766 a meeting had been held in Bradford to raise subscriptions for surveying. John Hustler, a Bradford merchant, was the leading promoter, while John Longbottom, a pupil of Smeaton's, and a native of Halifax, was the engineer who first suggested the scheme.[19] For a time the Leeds & Liverpool project had to contend with opposition from a rival plan to achieve east-west navigation by connecting the Calder and Mersey-Irwell navigations by a canal from Manchester to Sowerby Bridge. This, however, was shelved,[20] and a subsequent scheme for a York to Preston canal was short-lived.

Obviously such a canal as the Leeds & Liverpool could not be carried out without Lancashire participation and assistance. As soon as Longbottom has surveyed a line down the Ribble valley, and then through the Ormskirk district to Liverpool, "a committee of 14 gentlemen, of the said County of York, was appointed in order to obtain surveys and estimates for the prosecution of the said plan, and to meet and confer with the gentlemen of Lancashire".[21] Nevertheless, even before this conference was held at Preston on 1 March 1768, the first hint of possible dissension had already been given, for on 25 February the Yorkshire subscribers resolved to call in none of their subscriptions until the Lancashire men had paid in as much as had been raised in Yorkshire. Longbottom's survey was not adopted without question, and in fact a second survey was made. Liverpool Corporation, which had already subscribed £200 for the preliminary expenses, now agreed to pay another £50 towards the fresh survey, and seems to have insisted that Brindley be employed. The new survey was in fact made by Brindley and Robert Whitworth, and the former reported his conclusions to the

Liverpool promoters on 9 December 1768.[22] His report (which had already been given to the Yorkshire subscribers) was favourable to Longbottom's route, and the two committees met at Burnley ten days later to arrange an application for Parliamentary powers. The formal taking of subscriptions now began.[23]

Before Parliamentary approval could be obtained, however, a serious rift had taken place between the Lancashire and Yorkshire promoters. The Yorkshire subscribers, for their part, wanted the speediest communication with Liverpool consistent with cheapness. This involved omitting direct contact with Burnley and Blackburn in order to shorten the route. At the same time considerations of cost meant that the canal made a long detour through Burscough on its way into Liverpool, because a longer route on the level was cheaper than a short one with locks, and this deviation meant that Wigan, too, was by-passed. These features were naturally not welcome to the East Lancashire manufacturers and merchants whose towns were omitted from the route, nor were they acceptable to the citizens of Liverpool. From their point of view the canal would not serve their Lancashire hinterland so well, and, most important, the omission of the Wigan coalfield meant that it would not furnish an alternative supply of coal to that of the Sankey Canal. To the East Lancashire manufacturers it was cold comfort to be told that they could build branches from their towns to join with the canal, while it was misleading to suggest to the Liverpool public, as one Leeds & Liverpool propagandist attempted, that Lathom, Newborough, Dalton, and Skelmersdale were likely to provide all the coal which the town needed.[24]

As early as January 1769 the Liverpool Conversation Club was debating the need for a more southerly course for the canal "south of Ormskirk, through mines of coal, quarries of flag, slate, [limestone] etc. [rather] than north of Ormskirk, round by the sea, through barren land etc." If this were not feasible, then some members thought it could be joined at an unspecified point with the Sankey Canal, and thence connected with the Staffordshire Canal, which it was proposed to carry over the Mersey at Runcorn Gap by means of an aqueduct. While the junction with the Sankey was no doubt suggested by some interested proprietor or coal owner, the project of carrying the Trent and Mersey Canal over the latter river was, we have seen, already under discussion, and Brindley had been consulted upon it. So interested were the attenders at the Conversation Club that they advocated a survey.[25]

At a meeting of the Leeds Canal proprietors in Yorkshire on 14 June 1769 the Lancashire opposition was found to be so serious that Longbottom was sent off to see Brindley, "some of the Lancashire gentlemen being so warm for almost an entire alteration of the proposed canal through that county . . . we earnestly desire you will as expeditiously as possible settle with him the proper means of satisfying them how far it is eligible by a view of it or otherwise."[26]

The line Brindley was required to check was one surveyed on behalf of the Liverpool subscribers by John Eyes and Richard Melling, two Lancashire surveyors of some repute. Whitworth seems to have done the detailed work for Brindley, and claimed to find a major error in the Lancashire surveyors' level near Burnley. Hustler seized on this with some relief: "It is worth considering the impropriety of employing people so incapable the undertaking, or placing the least dependence upon their reports."[27] However, the Liverpool promoters initiated yet another survey, this time by the Liverpool surveyor P. P. Burdett. At this stage there was a distinct danger that the promoters on either side of the Pennines might break into mutually incompatible groups. Hustler smoothed matters over for the time being by agreeing that the Lancashire surveys should be completed, and the two lines then be examined by Brindley.

At the end of 1769 the Liverpool subscribers met at the Golden Lion in Dale Street to consider the Burdett route.[28] They informed the Yorkshiremen that they could not fall in with Longbottom's scheme, "having considered the impossibility that a committee of gentlemen in Yorkshire, be they ever so attentive, should be able to carry into execution a line of such an extent, and likewise being apprehensive that two committees might be impeded by a difference of opinion". They would therefore make themselves fully responsible for the canal on the Lancashire side, "and this is to be understood as a total cessation of all connections and a withdrawing of our subscriptions". Eighty-two signatures were then appended.[29] But the meeting of Leeds & Liverpool proprietors at Burnley a few days later rejected Burdett's plan, adopted Longbottom's, and decided to press on with the application to Parliament for the necessary Act.[30] The affairs of the canal had now reached a crisis. However, fresh deputations were sent to the Liverpool promoters and the Corporation of Liverpool, matters were somehow smoothed over, and many of the Liverpool dissentients again entered the fold, their only known gain from their opposition being the declaration that the Lancashire end from Liverpool would be proceeded with as fast as the Yorkshire one, a stipulation which, it might be thought, could have been obtained with much less ado. Strangely enough, it seems to have been Burdett's more southerly line, or an approximation to it, which was in fact built, after the long break in the construction of the central stretch of the canal between 1777 and 1790. The reasons advanced for giving the preference to the Longbottom line can scarcely have been convincing; while it may in fact have been the cheaper to build, its "admitting of a more easy communication with the northern parts of Lancashire and with Westmoreland"[31] meant that it was taking the route which would bring less traffic. The reasons for the collapse of the Liverpool opposition must remain obscure, but it is possible that one consideration leading the Leeds & Liverpool proprietors to close their ranks may have been the emergence of a new scheme of canal communication based on Liverpool.

By May the Leeds & Liverpool Canal Bill had become law,[32] and by November work was proceeding on the Newborough to Liverpool section.[33] But in the previous September the support of the Liverpool public had been sought for another scheme. This was to carry the Staffordshire Canal over the Mersey as had already been suggested, but, in this case, to join it to a branch starting near Duke Street in Liverpool, passing round the east side of the town, and connecting with the Leeds & Liverpool Canal at North Ladies Walk. The section between the Mersey aqueduct and Duke Street could be built through Appleton and Cronton and the Tarbock coal mines, and then enter Liverpool by Halewood and Toxteth. Once this was completed, it was suggested that an extension could be made to the coal district of Whiston and Prescot, with possibilities of further development through Ashton-in-Makerfield towards Bolton, Bury, Rochdale, and Halifax. This, its promoters said, would be a better scheme than that originally proposed of making a low-level branch from Runcorn to Liverpool, which could not have been connected with the Leeds & Liverpool and could not have been easily taken into the Whiston area, nor thence, if so desired, connected with a wide area of South Lancashire.[34]

The Liverpool Canal thus seems to have been something of a composite scheme, joining together various plans already broached. It would incorporate a footbridge as part of the aqueduct, and would enable passengers to get from Cheshire to Lancashire far more easily, and indeed passenger tolls would largely recoup the cost of the bridge's construction. It would carry the Staffordshire Canal across the Mersey, as had already been proposed. It would thus join two great canals, the Trent & Mersey and the Leeds & Liverpool, so that vessels could pass from one to the other. The course of the canal behind Liverpool would mean that there would be plenty of space for building and warehousing upon it and the Corporation could take the opportunity to establish the town markets in this relatively unoccupied district. The town's acute water shortage for industrial and other purposes would be relieved by building this virtual reservoir at its back and at a higher level. Last, but not least, it would be easy to extend the Runcorn end of the canal so as to communicate with the coalfield. As time went on it was this last part of the plan which aroused the greatest enthusiasm and altered the original idea almost out of recognition.

A week after this original advertisement in the newspapers, the projectors hastened to point out that their present subscription was only for the purposes of a survey, but that such subscriptions would establish a preference for any eventual share issue if the plan was found "eligible and advantageous".[35] The next week it was announced that the promoters of the scheme to join the two great canals "and for a branch out of ditto to Prescott, Bolton, &c." had met with such encouragement that there was to be a subscribers' meeting at Pontack's coffee house on 23 October. A president and committee would be chosen for the ensuing month to carry forward the plans and estimate the cost of the canal.[36]

When the meeting was reported an interesting addition had been made to the project, for the branch was approved of as running to "Prescot, Wigan, Bolton, &c."; there had clearly been pressure to put the Wigan coalfield on the route. Surveyors had been appointed to work out the canal's course, and three of the committee (of twenty) had been deputed to wait upon the Duke of Bridgewater to "know his pleasure respecting their design of going over at Runcorn to join his Canal".[37]

Whether unexpected difficulties were met, or whether the support shown by the subscription books was insufficient, is not clear, but the Liverpool press is silent upon the subject for nearly ten months. It was not until 16 August 1771 that a general meeting of those interested in the new canal "from Liverpool to Wigan, &c." was advertised for the 21st of the month.[38] A writer a few weeks later welcomed the new interest shown in the project—"the same laudable zeal for the public good seems reviving amongst us. . . From what I have heard and understand of the present scheme for joining the Staffordshire and Leeds Canals, it seems to be founded on the most generous principles, calculated for the general good of the country, and the trade of this port; on this account alone it merits the protection of the public, and therefore it may be hoped that no *selfish* principles or *private* views can wish the least opposition to its success."[39] Obviously the new canal was expected to arouse antagonism.

Preliminary plans were by this time available to the public, and an "inspired" letter to *Gore's Advertiser* the following week gave details of the route. Starting from Park Mills on the south side of Liverpool, it would run "under Sion Hill and Mount Pleasant, above the Infirmary, to the Loggerheads, through Everton Vale and Kirkdale on the south of Walton, Derby and Huyton, through the principal coal mines in Whiston, under Cronton to Appleton, upon one level; from thence to fall down by locks to the Runcorn Gap". The junction with the Leeds canal would be by a branch "from the Loggerheads to the North Ladies Walk", and this could be connected with the Mersey either by locks down to the river from "the north end of the town" or by wagon ways. The idea of taking the canal through Prescot and Knowsley had now gone, and instead a short branch was suggested to the Prescot collieries. The inland course of the canal was also somewhat revised: "from Appleton the same level will carry the canal south of Farnworth and Bold, cross the Sankey Brook by an aqueduct near Penkford Bridge north of Newton, near Ashton by Lowton School thro' the coal mines near Hedge Green, Bamfurlong &c., by Platt Bridge to near Bosden Bridge, there to join the intended canal from Kendal, Lancaster, Preston and Chorley, to his Grace the Duke of Bridgewater's canal at Worsley".[40] This branch would serve the Haigh and Arley collieries. From Bosden Bridge the canal could be branched to Wigan, and thence by a few locks to the Douglas Navigation. Thereabouts a junction could easily be made with the Leeds canal. This inland arm of the projected canal would open up about thirty miles of waterway into the mining and manufacturing district of south Lancashire; its southward sweep would complement the northward sweep of the Leeds canal and would greatly assist in opening up the Liverpool hinterland. Once again, as in the previous year, the advantages of the Runcorn bridge and the immediate communication it would open up with the growing canal system of Cheshire and the Midlands were rehearsed, as well as those of the canal in the vicinity of Liverpool.

Towards the end of this long letter, the likely source of opposition emerged, and an attempt was made to placate the Sankey Proprietors. When their vessels were detained by wind or tide in the Mersey, they could send on their coal to Liverpool by the new canal, "with certainty and expedition, at as small a charge as coming down the river, and receive their full tonnage as at present". They could also make a branch through the northern part of modern St Helens to join the new canal at some unspecified point, and thence bring their coal to Liverpool along a lockless and cheaper navigation.[41]

The Liverpool South Canal scheme seems to have fallen on more attentive ears than it had done a year before, probably because of the efforts made to turn it into a coal-carrying waterway, for it must soon have become apparent in Liverpool that the Newborough section of the Leeds canal would do little to break the Sankey coal monopoly. Subscription books for the Liverpool Canal were again opened, and it was further announced that it would not only pass near Wigan but also near Chorley, to join the Leeds Canal near Cuerdon.[42]

It is at this point that canal matters come to occupy a prominent place in the Liverpool newspapers over a period of some months. While the Lancaster Canal to join the Leeds & Liverpool was being brought to public attention, Hustler was writing to the Liverpool press to give a rosy picture of the progress of the Leeds & Liverpool, suggesting that it would only take a few years to complete, and that it would not exceed, and might even be cheaper than, the estimate.[43] In both suppositions time was to prove him grotesquely wrong. Some of the correspondence in the newspapers of the time is quaint rather than informative, but on 1 November 1771 a letter from one "Candidus" not only provides valuable information, but firmly ties the canal politics of Liverpool to the coal issue. Opening with the generalization that "such is the depravity of human inclination, so great the influence of private interest, that few public designs escape the malignity of opposition, from some selfish principles, or private views", the writer made a hard-hitting attack on the Sankey proprietors. He outlined the technical deficiencies of their canal, its disadvantages in falling into the tideway, and the fact that coal prices had risen rather than fallen during the period of its operation. But his chief target was a scheme for an opposition branch which the Sankey proprietors were now advancing, which was to run from Winwick Pond on their canal, through Newton, Lowton, Abram, and Goldbourne to Bryn Moss in Ashton, and thence by branches to Ince and Pemberton. This scheme was designed, as was the Liverpool Canal, to tap the coalfields south of Wigan, but "Candidus" did not find it equally acceptable. Though the Sankey proprietors proposed making the new branch into a sort of ship canal, so that vessels of 80 to 100 tons could reach Wigan, it would still open into the Mersey tideway with all its disadvantages, "so that the poor inhabitants of Liverpool may still starve in winter, with joyful expectations of having larger morsels in summer". The Sankey party claimed that with this extension their canal could itself fulfil all Liverpool's needs, but there was "not a word of the price". On the present length of the Sankey vessels only made two or three trips down to Liverpool and back per month, but the Liverpool Canal boats would be able to reach Ashton or Hindley in about twelve hours from Liverpool. The Liverpool Canal, too, would provide a better coal wharf at the port than that of the Sankey proprietors, for their place of delivery was "so excessively dirty, there's no seeing the quality or measure before they are delivered, and no stocks ever kept on hand". Against the claims of the Sankey proprietors that the crossing of their waterway by the proposed canal would obstruct its navigation, it was pointed out that the Irwell Navigation had been crossed by the Barton Aqueduct without any interference being suffered. In fact the Sankey proprietors would have little to complain of; the Liverpool Canal carefully avoided poaching on their present coal preserves, while against their claim that they had had their extension in mind since 1768 it could be argued that they had done nothing about it, while the Liverpool Canal, whose practicability they had denied, was now before the public.[44]

On 8 November another writer supported the arguments of "Candidus" at some length. He declared himself to be a Leeds canal proprietor, but confessed that a great error had been made in planning the wide northerly sweep through Newborough and Halsall. The Liverpool Canal, on the contrary, would run through many mining areas which "are certainly the most desirable and most advantageous object, for the town and trade of Liverpool, of any [canals] yet put into execution". He agreed that the various canal projects should avoid needless competition, so as not to "end in the destruction or loss of the different adventures". But, "as there cannot possibly be an overstock of the article of coals to such a trading port as Liverpool, private interest ought always to give way to public utility". All the local navigation and canal projects had approached Parliament for powers on the basis of their value to the public, but coal was now 15 to 20 per cent dearer in Liverpool than when the Sankey undertakers applied to the legislature, while the owners were making an annual dividend of 20 per cent and over. He believed the new Sankey extension scheme to be a "pretext", introduced "when they see the people of Liverpool will no longer groan under their yoke".[45]

The support which the Liverpool Canal had now gained in the town can be seen from the mayor's calling a public meeting in the Exchange to discuss "an intended navigable canal . . . to or near Wigan", when a full gathering approved it. A former mayor, John Sparling, as one of the principal managers of the scheme, had opened the Liverpool subscription, and books were also opened in Wigan, Warrington, and Chorley.[46] While the Liverpool Canal men were striving to raise the £60,000 they estimated as required, the Sankey proprietors, probably as a piece of camouflage, took in subscriptions for their extension, insisting both that the project would be innocent of private monopoly, and that the money was raised in the record time of a quarter of an hour.[47]

It is at this point that canal politics become murky. On 21 December 1771 Hustler was directed by the Leeds & Liverpool Committee to write to Jonathan Blundell and Mr Earle of Liverpool to thank them for their services as intermediaries in buying from Mr Alexander Leigh of Wigan his majority share in the Douglas Navigation. At the same time the Committee ordered Longbottom and others to examine the collieries around Wigan, to pay Leigh his first one-quarter instalment of the Douglas purchase money, and then to go on to Liverpool, "to consult with proper persons there about supplying the town with water from the Leeds & Liverpool Canal and also about taking such steps as shall appear most advisable for opposing and rendering ineffectual the scheme for an Act for making a . . . canal from Wigan to Liverpool".[48] Early next month the emissaries reported favourably on their mission. The banks of the Douglas were well lined with coal measures. One of the party, Mr Hardcastle, had had the opportunity of meeting the Corporation and inhabitants of Wigan, and, in return for their support, "proposed that the Company would engage to take no more than one shilling per ton for the navigating of merchandise from Wigan" to the junction with the Leeds canal. Hardcastle had then proceeded to Liverpool, where he, Earle, and Blundell, for the Leeds-Liverpool proprietors, had signed a pact with Dr Boscough and Scrope Calcott, Esq., for the Sankey proprietors, agreeing that the proprietors of the Leeds & Liverpool Canal should join them and bear one-third of the expenses of opposing the Liverpool-Wigan Canal, "the landowners bearing one third, and the proprietors of the Sankey Navigation the remaining third part".[49]

The Holt-Gregson Papers are thus entirely justified in stating that the purchase of the Douglas Navigation by the Leeds Canal was made "as one means of preventing" the Liverpool Canal's being sanctioned and reaching the colliery districts.[50] The blow was indeed a damaging one, and the Liverpool Canal supporters immediately wrote to the press in an endeavour to prove that their scheme was still valid. Its course "thro' a chain of coal mines 30 miles or more long" could not be improved upon; many of the Sankey and Leeds & Liverpool subscribers believed that the Liverpool Canal scheme was the most promising and had given it their subscriptions; and £48,000 of the £60,000 needed was now guaranteed. If the Sankey and Leeds canals could, as they claimed, provide Liverpool with coal as cheaply as the new venture, "let the poor, the mechanic, the trader, the merchant rejoice at the hope of so valuable an acquisition"—all the coal brought could be consumed. There was thus no need for "jarring interests" or "undue measures by any of the adventurers".[51]

By the middle of December surveying for the Liverpool Canal was complete, the Parliamentary plan was in preparation, and negotiations with the relevant landowners were about to open. But there were "frequent and reiterated attempts" to obstruct this "glorious undertaking".[52] Many came from the Sankey proprietors. Not only had they "cooked up a subscription among such of their choice friends as they think worthy of their society" for their extension scheme, but they had attempted, with fairer words, to get the Liverpool Canal supporters to run their canal into the Sankey Navigation, thus using the existing canal for part of their course, though this would have meant an immense array of locks.[53] This advance refused, they spread rumours among the people of the town that the Liverpool Canal, by running behind and above the built-up area, would cause flooding, while the landowners were told that their lands would be cut up unnecessarily.[54]

Forces were now gathered for the Parliamentary struggle. The Liverpool Canal reached its £60,000 subscription target, but continued to entice the subscriptions of landowners[55] in order to counteract the adverse canvassing of that body by their rivals. There was some last-minute discussion of their past dissensions between those Leeds & Liverpool proprietors who had crossed into the Liverpool Canal camp and others who had returned after the Douglas purchase; and the failure of the Yorkshire party under Hustler to provide the best line through Lancashire was once again criticized.[56]

In January 1772 the Liverpool Canal projectors petitioned Parliament for leave to bring in a Bill to make a canal from Liverpool to Wigan—the Runcorn Bridge plan seems to have been abandoned, as does the idea of continuing onwards from Wigan towards Chorley. In the next few weeks, however, they had to face counter-petitions from the Sankey proprietors, the Lancashire landowners and the Leeds & Liverpool proprietors.[57] The former claimed that their own canal supplied Liverpool efficiently, and that between the Leeds and Bridgewater canals and their own there was all the transport the district required. They also objected to the aqueduct which would be necessary to carry the Liverpool Canal over their own. The landowners, headed by Lord Derby, declared that additional transport was not needed, as the turnpike brought plenty of coal from Prescot and Whiston—an argument which was at least twenty years superseded. Finally the Leeds & Liverpool proprietors contented themselves with emphasizing the adequacy of the canals then building, and the merits of their own junction with the Douglas Navigation.

By the chance preservation of a series of letters between the Liverpool Canal promoters who acted as lobbyists in London and the main committee back in Liverpool, we are able to see the causes of their Parliamentary defeat in some detail. The deputation managed to get the support of Mr Gascoigne[58] and of Lord Sefton,[59] but as late as 25 January the plans prepared by Messrs Yeoman and Everard[60] had not yet reached London. They were also in need of someone to give expert evidence on the coal trade, and desired that Mr Richard Melling of Wigan, a noted coal viewer, should be sent up. There was little generosity between the different industrial regions at that time, and it was quickly discovered that "the very apprehension of Liverpool becoming a coal-port, will certainly bring upon us the opposition of Cumberland and Northumberland".[61] As a counter to this the lobbyists tried to discover how many corn ships from the south of England to Liverpool returned in ballast for lack of a back-cargo, so as to obtain southern support for the canal both on the grounds of breaking the existing monopoly of Cumberland and the North-East and of assisting southern shippers.[62] Next the rather amusing claim of the Sankey owners in their petition concerning "the great use of their navigation in raising seamen for His Majesty's Navy (a very popular idea) . . . must either be circumstantially confuted or it will carry great weight against us".[63] Rebuttal could only be made by showing that Sankey flats carrying coal for the coal-masters were but a small part of the traffic, much being undertaken by the salt-boilers' own flats, which would, whatever the fate of the Sankey, continue in use on the waterway which offered them the best access to the coalfields.

When eventually a meeting was held with Lord Sefton, Sir William Meredith, and other Members at the Salopian coffee-house, the Parliamentary prospects were found to be far from bright. Lists were produced of the landowners through whose lands the canal was to run, but it was found that two-thirds of their number opposed the scheme, though they were estimated to hold less than half the acreage to be traversed. When it was considered that the House had "a natural jealousy to all navigations", the likelihood of a Parliament of landowners passing the Bill was small. Even its warmest supporters could not be too open in their advocacy in such circumstances—"Lord Sefton said that as many considerable land owners were in opposition, although he himself did consent and in his private capacity did promote the scheme, that he could not in his public one solicit the votes of others in Parliament to support it". Since the ensuing session was expected to be short, the members who were friendly to the plan suggested that the promoters should conciliate some of the landowners and apply again later. But here the Liverpool Canal promoters were in a cleft stick, for they knew that the Leeds canal would soon be complete from Liverpool to the Douglas, and that the Leeds & Liverpool and Sankey proprietors would then collaborate in furnishing Liverpool with a cheap and plentiful supply of coal for just as long as was required to convince Parliament of the inutility of an additional canal. They therefore decided that it had better be now or never; whatever its chances, they would submit their petition to the examination of a Commons Committee, and they would endeavour to "make friends by our merits that we could not make by our connections".[64]

On 27 February the scheme was before committee for three hours, and the surveyor, Mr Everard, gave his evidence very satisfactorily, Mr Gascoigne giving him all the help he could. Between them they were able to show that "it is not that unnecessary, chimerical scheme that our opponents have represented", but their friends in the House were still of opinion that they should, after exhibiting the merits of their plan, temporarily withdraw it and conciliate the landowners. This was done at the next session of the Committee, when Gascoigne and Meredith executed a graceful withdrawal on behalf of the promoters until such time as the scheme could be satisfactorily explained to the landowners. The deputation had to return to Liverpool empty-handed, "hoping that what we have done in this business (notwithstanding our failing in success) will have the approbation of all the parties concerned".[65]

Apart from a solitary mention, nothing further is heard of the Liverpool Canal scheme in its original form.[66] But in 1794, when trying to force the hand of the Leeds & Liverpool proprietors, and induce them to charge lower rates, the Canal Committee of the Common Council used a similar scheme as a threat. This was at the time when the deviation to Burdett's approximate line was being applied for in Parliament. While this was greatly to the advantage of Liverpool, the proprietors threatened to abandon the alteration if the Common Council pursued its intention of demanding cheaper rates for Wigan coal while the Bill was before the Houses. The Canal Committee issued the counter-threat that if the deviation plan were withdrawn the Council would "consider of the Propriety of calling a General Meeting of the Inhabitants to take into consideration the expediency of making a new canal from this town to Wigan by a nearer line, which has been suggested to this Committee, as a measure likely to secure an abundant, cheap, and regular supply of the article of coals to the public at large".[67] Perhaps the last occasion on which an ambitious scheme for a canal through south Lancashire emerged was in a plan of 1825 by Francis Giles, civil engineer, of London. He would not only have made an aqueduct and roadway across Runcorn Gap, but have linked the Bridgewater, Sankey, and Mersey-Irwell systems.[68] Thereafter canals were a dead issue—in Lancashire the Railway Age had entered upon its vigorous life.

While the Liverpool Canal never came to anything, its story is worth telling because it teaches us much about the Liverpool attitude to canal construction at the most important period. While they were willing to assist the most ambitious cross-country canal projects, the Corporation and citizens of Liverpool placed the first emphasis on the efficiency of canals as coal carriers, and they were correct in regarding them as of more importance as short-distance rather than long-distance hauliers. The failure of the Leeds & Liverpool as a short-range coal supplier to Liverpool nearly caused the wreck of the scheme, and it was only after it had been threatened by a dangerous rival, and had made belated arrangements to connect with the Wigan coalfield, that it made good its privilege of being the only canal across Lancashire to Liverpool. The figures of coal carried from Wigan to Liverpool[69] in the first decades of the existence of this section of the canal are sufficient commentary on the value of the opening-up of that coalfield. On the other hand the Liverpool Canal must remain an intriguing might-have-been: while its chances of success were diminished by the many alterations the plan underwent, by the failure to engage a canal surveyor or engineer of national repute,[70] and by the forces of vested interest and reaction which marshalled themselves against it, it seems to have been an idea of real merit, likely to open up the South Lancashire hinterland of Liverpool more effectively than the Leeds & Liverpool Canal in its unrevised form would have done.[71] It failed, but its very proposal and discussion led to the satisfaction of its main object—the improvement of Liverpool's coal supply. The preoccupation of Liverpool's citizens with this coal question makes us realize how serious Merseyside's persistent fuel crisis seemed to contemporaries, and it partly explains why waterways so advantageous to the port as the Trent & Mersey and the Bridgewater canals had proportionately so little space devoted to them in the controversies of the time. Of course the Bridgewater scheme was in a sense above criticism—there was a recognition that the designs of the Canal Duke were inspired by something higher than profit, while in a concern owned by a single aristocratic proprietor, there was no possibility of shareholders' quarrels occurring, or being aired in print.

The Liverpool canal controversies of 1769–72 are only an episode, but an episode to be set against the long perspective of Merseyside's transport history. In few regions of Britain can the story of inland transport over the past 250 years be more fascinating. There is the first dock in the outports in 1710; the first canal of the Industrial Revolution; the speed with which Liverpool was connected to the canal system; the Rainhill trials and the Liverpool & Manchester Railway; the influence of Liverpool capital in the creation of a national railway system; the Manchester Ship Canal controversies; the Mersey Tunnel in our own day. Even so, this leaves out of the question sea transport and all concerned with it, ships and docks, shipowners and sailors. To the historian of Merseyside the economics of transport must always have a fundamental importance.

A NOTE ON THE STAFFORDSHIRE CANAL EXTENSION PROJECT

The low-level scheme referred to on page 163 was presumably that of Brindley, which had been under intermittent discussion since the end of 1765. A Liverpool group headed by John Tarleton had been interested in the extension of the Bridgewater Canal (and later also of the Staffordshire Canal after their junction had been decided) over the Mersey by an aqueduct at Runcorn. However, they do not seem to have pressed it with sufficient zeal, though there was considerable activity in the project at the end of 1768 and the beginning of 1769. Brindley seems to have been thoroughly in favour of the plan. "I find he hath now (Nov. 1768) the same opinion of the Runcorn Aqueduct which he often expressed to me and others two or three years ago—That there is no doubt of the practicability of its being made nor of its answering to the undertakers." But Brindley's design seems to have been for a combined road bridge and aqueduct, with a rather narrow waterway; when a decision was made to widen certain sections of the Staffordshire Canal and to make that canal and its Bridgewater connection with the Mersey capable of taking a wider vessel than the projected aqueduct could, Brindley began to lose interest. In his opinion, if the Liverpool gentlemen did not push their scheme ahead fast enough to make it worth the while of the Bridgewater and Staffordshire Canal promoters to abandon their widening plans in exchange for a direct communication with Liverpool, it might as well be dropped. The Liverpool Corporation threatened at one stage to prevent the widening of the Staffordshire Canal by taking legal action—a rare unworthy episode to be set against the otherwise enlightened attitude of the Corporation towards transport improvement. Tarleton thus lost an opportunity of glory: "My best respects to Mr Tarleton, we will be preparing some Etruscan Earth to make his Statue with that of Brindley L'Grand to grace the most conspicuous part of this tremendous fabric."[72] Smiles suggests that the Duke of Bridgewater's financial difficulties provide the reason for the failure to build the aqueduct and the continuation of the canal into Liverpool. But Picton's extract from the Town Books makes it plain that both the proprietors of the Staffordshire Canal and the Corporation of Liverpool were manifesting a keen interest in the scheme in 1768, and that it was at the Corporation's request that Brindley surveyed the course of the canal between Runcorn and Liverpool and gave in his estimates for both aqueduct and canal. The Duke was merely approached for his consent that the two waterways be joined at Runcorn. One source would indicate that the survey was made.[73]

NOTES

The writer wishes to express his thanks to Mr Gordon Benzie for his assistance in planning the accompanying map, to Mr R. W. G. Bryant, Dr W. H. Chaloner, Mr C. Hadfield, and Dr S. B. Saul for advice and comment, and to the staff of the Liverpool Record Office.

1. Liverpool Record Office, Holt-Gregson Papers, 5/93, 95.

2. The Sankey was complete over its main course in 1757. The Liverpool-Wigan section of the Leeds & Liverpool was opened in October 1774, though that canal as a whole was not finished until 1816. The Bridgewater and Trent & Mersey schemes were under construction during the period of the Liverpool controversies, the former being completed in 1776 and the latter in 1777. The Duke of Bridgewater, however, was able to sell coal in Liverpool as early as 1774.

3. T. C. Barker and J. R. Harris, *A Merseyside Town in the Industrial Revolution* (1954), chapters II to V: T. C. Barker, "The Sankey Navigation", *Trans. Historic Soc. of Lancs & Cheshire*, 100 (1948), 121 *seq*.

4. Barker and Harris, *op. cit.*, 15–23. 5. Holt-Gregson Papers, 10/177.

6. Liverpool's coal-consuming industries at this period included breweries, potteries, sugar refineries, glasshouses, a salt refinery, and a copper smelting concern.

7. *Williamson's Lpl. Advtr.*, 12 Feb. 1762. 8. *Ibid.*

9. *Ibid.*, 8 May 1761, 9 Dec. 1774; *Gore's Lpl. Genl. Advtr.*, 1 Nov. 1771.

10. *Williamson's Lpl. Advtr.*, 8 Jan., 12 Feb. 1762.

11. *Ibid.*, 14 Jan. 1774, quoted Barker and Harris, *op. cit.*, 47. This bears out Smiles's observations on the Duke's consideration for his poorer Manchester customers: *Lives of the Engineers: James Brindley and the Early Engineers* (1874), 229.

12. Barker and Harris, *op. cit.*, chapter III, *passim*.

13. Canal boats, in the case of the Sankey almost invariably equipped with sails.

14. Barker and Harris, *op. cit.*, 41–2, 45. 15. *Ibid.*, 28–9.

16. J. A. Picton, *Liverpool Municipal Archives and Records (1700–1835)* (1907), 243.

17. *Ibid.* J. Phillips, *A General History of Inland Navigation* (4th ed., 1803), 115, states that Liverpool Corporation supported a Trent & Mersey survey by Taylor and Eyes as early as 1755. For John Eyes see R. Stewart Brown, "Maps and Plans of Liverpool and District by the Eyes family of Surveyors", who also mentions this survey: *Trans. Hist. Soc. of Lancs. & Cheshire*, LXII (1911), 143 *seq*.

18. *Williamson's Lpl. Advtr.*, 28 Feb. 1766. John Rylands Library, Wedgwood Corr., Wedgwood to Bentley, 5 Dec. 1765: "Do you mean that His Grace's Canal (Bridgewater) should not proceed as now laid down in the plan . . . but should not (*sic*) cross the Mersey perhaps a little above Warrington and be carried on to Liverpool and so not fall into the Mersey at all? Where do you mean the Aqueduct to be?"

Notes

19. H. F. Killick, "Notes on the Early History of the Leeds and Liverpool Canal" in *Bradford Antiquary*, n.s., II (1897), 175–9; J. Priestley, *Historical Account of the Navigable Rivers, Canals and Railways* (1831), 385–6; *Williamson's Lpl. Advtr.*, 9 Oct. 1767.

20. A canal between these places was constructed much later as the Rochdale Canal.

21. *Williamson's Lpl. Advtr.*, 19 Feb. 1768.

22. Killick, *op. cit.*, 185–7; J. A. Picton, *Liverpool Municipal Archives and Records* (1907), 244; *Williamson's Lpl. Advtr.*, 10 Feb. 1769.

23. *Williamson's Lpl. Advtr.*, 20 Jan. 1769.

24. Killick, *op. cit.*, 190 *seq.*; *Williamson's Lpl. Advtr.*, 10 Feb. 1769.

25. *Williamson's Lpl. Advtr.*, 20 Jan. 1769. 26. Killick, *op. cit.*, 191. 27. *Ibid.*, 192.

28. *Lpl. Genl. Advtr.*, 1 Dec. 1769. Burdett was an artist, engraver, and the inventor of an aquatint process (over which he quarelled with Wedgwood), as well as a surveyor.

29. Killick, *op. cit.*, 196. 30. *Lpl. Genl. Advtr.*, 5 Jan. 1770.

31. Killick, *op. cit.*, 198; *Lpl. Genl. Advtr.*, 5 Jan. 1770. 32. 10 Geo. III, cap. 114.

33. *Lpl. Genl. Advtr.*, 9 Nov. 1770. 34. *Ibid.*, 28 Sept. 1770.

35. *Ibid.*, 5 Oct. 1770. 36. *Ibid.*, 12 Oct. 1770. 37. *Ibid.*, 26 Oct. 1770.

38. *Ibid.*, 16 Aug. 1770. 39. *Ibid.*, 6 Sept. 1771.

40. Whitworth, on Brindley's behalf, surveyed part of the Lancaster Canal in 1772, but no Act was obtained till 1792. For its course when completed see Priestley, *op. cit.*, 372–7.

41. *Lpl. Genl. Advtr.*, 13 Sept. 1771. 42. *Ibid.*, 11 Oct. 1771. 43. *Ibid.*, 18 Oct. 1771.

44. *Ibid.*, 1 Nov. 1771. Extracts are given in Barker and Harris, *op. cit.*, 39–40.

45. *Lpl. Genl. Advtr.*, 8 Nov. 1771. 46. *Ibid.* 47. *Ibid.*, 8 and 15 Nov. 1771.

48. Killick, *op. cit.*, 219 *seq.*

49. Killick, *op. cit.*, 221 *seq.* "Dr Boscough" may be an error for Dr John Bostock, who married a sister of Nicholas Ashton, principal proprietor of the Sankey Navigation, in 1771. "Calcott" is usually written "Colquitt".

50. Holt-Gregson Papers, 10/215.

51. *Lpl. Genl. Advtr.*, 6 Dec. 1771. The same letter gives the Committee of the Liverpool Canal as John Tarleton, John Sparling, and James Gildart Esqrs., and Messrs Thomas Wickliffe, Edward Chaffers, Roger Fisher, and J. Hollingshead. Tarleton, Sparling, and Gildart were aldermen, all three were merchants, and Tarleton and Gildart both took part in the Africa trade. Tarleton was also a sugar baker, and owned estates in the West Indies. Gildart had an interest in the coal and salt trade. Chaffers was the son of the pioneer of the Liverpool porcelain manufacture. Fisher was in business as a shipwright. Hollingshead styled himself "gentleman".

52. *Lpl. Genl. Advtr.*, 13 Dec. 1771.

53. Liverpool Record Office, Binns Collection 31/212. Public letter of 19 Nov. 1771.

54. *Lpl. Genl. Advtr.*, 21 Feb. 1772. 55. *Ibid.*, 20 Dec. 1771. 56. *Ibid.*, 27 Jan., 7 Feb. 1772.

57. *Commons Journals*, XXXIII, 22 Jan., 10 Feb. 1772.

58. Sir William Meredith, Bart., and Richard Pennant were M.P.s for Liverpool at this time; Bamber Gascoigne, Sen., was M.P. for Weobley; his son, Bamber Gascoigne, Jun., became M.P. for Liverpool in 1780.

59. An important landowner in Liverpool and district.

60. Thomas Yeoman was surveyor to the Stroudwater Navigation, 1774–9. See C. Hadfield, *Canals of Southern England* (1955), 61. William Everard (1723–92) had some reputation as an architect, designing among other buildings the uncompleted Liverpool Observatory. He was also founder of the Liverpool Library and writing-master to the Corporation. See P. Fleetwood-Hesketh, *Murray's Lancashire Architectural Guide* (1955), 79.

61. Wigan Public Library, Liverpool and Wigan Canal Letter Book; Deputation to Staniforth, 6 Feb. 1772.

62. *Ibid.*, Wycliffe to Dobson, 9 Feb. 1772. 63. *Ibid.*, Deputation to Committee, 11 Feb. 1772.

64. *Ibid.*, Deputation to Committee, undated letter (mid-Feb.) and 20 Feb. 1772.

65. *Ibid.*, Deputation to Committee, 27 Feb., 2 Mar. 1772.

Notes

66. *Lpl. Genl. Advtr.*, 4 Sept. 1772: asks for remaining payments on a second call of 15s. per cent to be made.

67. Lancashire Record Office, Moore Canal MSS., fol. 354. Notes of a meeting of the Corporation Canal Committee held in Exchange, 8 Feb. 1794.

68. *Ibid.*, fol. 592.

69. Holt-Gregson Papers, 10/215, give the following tonnages (22 cwt. to the ton):

1781	31,401	1784	70,555	1787	98,246	1790	137,790
1782	39,326	1785	85,138	1788	109,209	1791	145,616
1783	57,376	1786	91,249	1789	124,874		

By 1791, except for some deliveries to breweries, the Leeds Canal had largely replaced the Sankey as the supplier of Liverpool's fuel. This followed a price-war between the two canals after 1774, and a three-cornered tussle with the Bridgewater after 1776. See also *ibid.*, 19/39.

70. It may be worth noting that at least one of the supporters of the Liverpool-Wigan Canal had a dread of over-powerful individuals among the canal proprietors having too great an influence. He asked that "not less than half a dozen men, endued with common sense", should be asked to view the intended canal before application were made to "Mr. Smeaton, or any other Engineer".

71. That is, before the Douglas purchase. Later developments such as the further improvement of the Douglas, the arrangement to share part of the course of the Lancaster Canal, the Wigan-Worsley extension to join the Bridgewater system, lie outside the period under discussion.

72. Wedgwood Corr., Wedgwood to Tarleton, 12 Nov. 1768; Wedgwood to Bentley, 16 Jan. 1769 and 18 June 1769; Stronge to Wedgwood, 22 July 1769; *Lpl. Genl. Advtr.*, 21 Oct. 1768.

73. Smiles, *op. cit.*, 215; Picton, *op. cit.*, 244; J. Phillips, *General History of Inland Navigation* (4th ed., 1803), 110.

4 *Bradford coal, Craven limestone and the origins of the Leeds and Liverpool Canal, 1765–75*

G. FIRTH

For an industrial town of its size and importance, surprisingly little has been written about Bradford's pre-industrial development, in particular about the keenly contested debate on the relevance of transport change to economic growth. Several recent studies of transport history, and of canals in particular, have made but passing reference to the close association of the Leeds & Liverpool canal scheme with the town of Bradford. J. H. Farrington, for example, has commented, 'The canal was mainly Bradford-inspired in the early days and the main controlling committee was largely composed of Yorkshiremen for many years.'[1] This affinity has been reflected in the main secondary literature and probably originates in Joseph Priestley's *Historical Account of the Navigable Rivers, Canals, etc.* (1831).[2] Priestley's father was clerk of works to the Leeds & Liverpool Canal Company for almost fifty years. Bradford's strong ties with the project were also confirmed in 1900 by another employee, H. F. Killick, who was law clerk to the company.[3] Both writers emphasised the substantial Bradford influence on the affairs of the Leeds & Liverpool Canal from its beginnings in the 1760s. The purpose of this article is to examine why Bradford took such a close promotional interest in the canal and in so doing to reinforce the familar themes of the interrelating links between fuel, transport and industrial development.

I

Bradford lies in a basin-like depression on the eastern side of the Pennines, surrounded largely by hills of up to 1,200 ft above sea level. This valley, a branch of Airedale, stretches from the moorlands above Thornton to the river Aire at Shipley and in 1750 still warranted the appellation of Bradfordale. Cut off from the main regional arteries of transport and communication, Bradford had remained a small market township incapable of significant commercial and economic progress.[4] By 1850, however, it had developed into an important regional centre of road, navigational and railway communications and had undergone 'a remarkable change from its position as an economic backwater to that of a major manufacturing city owing much to its emergence as a man-made node . . .'.[5]

The primary stimulus to this transformation came in the second half of the eighteenth century with the growing regional demand for Bradford's relatively untapped mineral wealth. Coal mining was central to the history of the British industrial revolution, and in particular to the early industrialisation of the Bradford region. The town stands on the north-west corner of the Yorkshire coalfield, in which there are two distinct types of coal, the Black Bed and the more famous Better Bed coal. To the south of Bradford the latter seam had an average thickness of twenty-five inches. Its coal was described as a 'bright burning bituminous coal' which was sulphur-free and ideally suited to the iron trade. A third type, the Hard Bed coal, was close to the surface and was for many years an important source of domestic fuel.[6] As the town was located on the edge of the coalfield, the seams to the north began to peter out quickly and early mining schemes were easily exhausted.

Simple digging at the outcrops of coal had occurred since the fourteenth century, but by 1700 there was a proliferation of small-scale mining schemes to the south and south-west of Bradford. Local lords of the manor and leading estate owners in this area had a long tradition of non-agricultural involvement. As far back as 1568 Sir Richard Shireburn had profited from timber on his Esholt estates. The inferior wood he burnt in an iron bloomery at Esholt Wood.[7] By 1750 several of the leading Bradford gentry were actively taking up the role of industrial entrepreneur, involving themselves directly in capital investment and entrepreneurial risk. And the exchange trade of Bradford coal and Craven limestone was a major element in this activity.[8] Edward Leedes, of Low Moor and North Bierley, is a typical example of the Bradford gentry in these years. His activities in the field of mineral extraction were extensive and included an eighth share in a lead mine on Grassington Moor, in addition to numerous collieries at North Bierley and Wibsey.[9] By 1760 he had planned for his mineral estates something more than scratching at the surface coals. He had conceived an enormous colliery network linked by railroads, to provide for wider markets and even exports. There was talk of a 'great plan of improvement'. However, he failed in his attempt to raise capital and realise the full potential of this scheme: money was scarce, and the American war affected interest rates. In August 1781 he was declared bankrupt, but continued to believe in his plan. 'The small Wood Colliery,' he wrote, 'will bring in £3,000 a year if there was a Newcastle Waggonway made to Brigghouse and £1,000 a year if it was only made to Lightcliff.' He made greater claims for the large colliery: 'there is the manor of 500 acres of Common, Ironstone etc. Reversion of the £1,000 p.a. Colliery at the third year and which will bring in £10,000 a year and coal getting, two hundred years . . .'.[10]

Had Leedes continued to live the life of the landed squire, drawing his income from yearly rentals, there was every possibility that his family name would have survived. He chose, however, to risk borrowed capital in the exploitation of his subterranean estates. After 1775 he drew annually upon persons in London for large sums of money amounting to £20,000 and more.[11] His single-handed efforts to finance the collieries, as well as carry the burden of a heavily mortgaged estate, ultimately ruined him, and in August 1785 he took his own life. It has been said of Edward Leedes, 'To him the mineral wealth of his vast estates was an unknown quantity; he lacked the energy to reap the rich harvest which subsequently fell to the lot of the Low Moor Iron Company.'[12] This is an undeserved comment, for he was one of the many landowners who risked and lost everything in the pursuit of commercial and industrial success. Here, it is important to recognise that the ambition and foresight of Leedes, and his 'great plan', were built upon the growing regional trade of Bradford coal for Craven limestone.

The use of calcinated limestone for building purposes had been known for many years, and by 1700 lime was put to a wider use in the West Riding as a fertiliser.[13] Much of the limestone in mid-Airedale was found around Skipton and Bingley, where it was burnt in rectangular 'pye-kilnes' close to the quarries. Throughout the eighteenth century several Bradford colliery owners like Leedes increasingly moved their coal up the dale towards Skipton and in return received lime or limestone which was burnt in the limekilns at Bingley.[14] Edward Leedes was joined in this local trade by Abraham Balme of Bowling, John Hustler of Undercliffe, John Stanhope of Thornton, John Jarratt, John Preston and many others. These men took advantage of the increasing potential of coal as a domestic fuel where the transport system allowed, for the movement of large quantities of minerals exposed severe limitations in the region's existing transport network. Heavy but cheap commodities like coal and limestone could have their price raised to a disproportionate extent by transport costs. The carriage of minerals by packhorse through Airedale was painfully slow, inhibited output and sustained high prices.[15] In spite of such restraints the Bradford coal dealers carried on a sound business in the middle decades of the eighteenth century. The trade in coal and limestone was one of the most prominent features of Airedale's market economy at that time. The demand for coal as a domestic fuel, and the needs of Bradford farmers and builders for high-quality lime, strained the local transport system more than any other commercial traffic during this vital economic period.

II

As early as 1744 there had been a proposal to make the river Aire navigable from Inghay Bridge to Cottingley Bridge in order to accelerate the movement of coal between Bradford and Skipton. The final estimated cost of £5,000 proved prohibitive, and the scheme was dropped.[16] In consequence the Bradford colliery owners were entirely dependent upon the local road network. As the inter- and intra-regional commodity flows increased in volume in the eighteenth century the condition of West Yorkshire roads was called into question by their primary users, the gentry and merchants. Between 1734 and 1760 Bradford was drawn into a network of turnpike roads whose trustees included the familiar local names of Balme, Leedes, Hustler, Preston and Stanhope.[17] The close involvement in road improvements of the Bradford colliery owners did little to affect 'the prodigious expense of land carriage'. John Hustler, like the petitioners of 1744, acknowledged this as a primary problem in his propaganda pamphlet:

> The price of land carriage at present is on an average one shilling per mule per ton: the price proposed to be taken by water . . . for merchandise two pence per ton mile; for coals a penny half penny per ton per mile, or one eighth part of the land carriage . . .[18]

In a private survey of 1768 the Bradford canal promoters found that 46,880 horseloads of coals were consumed in the Skipton district at a cost of tenpence per load. It was calculated that the delivery of coal by navigation could be made at a maximum of eightpence per load and the cost of limestone could be cut by half. These preparatory calculations estimated an annual profit margin at £21,132, of which £7,200 would come from the carriage of merchandise and £13,932 (65 per cent) from coal and limestone.[19] As early as August 1764 there appeared a newspaper comment that:

> As the Rivers Aire and Ribble may be so easily joined at different places and rendered navigable between Leeds and Preston at an expense which gentlemen who have estates on the banks may readily supply it is thought proper to mention it to the public at this juncture.[20]

At that time the economic benefits of the Sankey brook and the Bridgewater Canal were generally known. Early canals tended to be local schemes, financed and organised by those who had a direct interest in the results.[21] The Leeds & Liverpool Canal was initially conceived to meet the specific needs of a small group of Bradford colliery owners, keen to cut costs and accelerate distribution to the traditional regional markets of the Yorkshire dales. Most secondary works have credited John Hustler, the Bradford woolstapler and coal merchant, with the original idea of a trans-Pennine canal. Two writers have gone so far as to dedicate their work to his memory.[22] Certainly the initiative came from within the cabal of Bradford colliery owners, but another candidate deserves mention in this context, namely John Stanhope of Horsforth and Thornton. Stanhope's career and those of his brothers, Walter and William, have been examined elsewhere.[23] As an attorney John had made a name for himself in the woollen cloth disputes of the 1760s, representing the case of the Leeds clothiers. Having inherited the Horsforth estate, he purchased the manor of Thornton in 1746. Stanhope then encouraged local men to lease his Thornton lands and invest in mining schemes. In June 1765 he received a letter from his nephew, Walter, who had recently visited the impressive Bridgewater Canal. Walter observed:

> We went in a pleasure boat down ye navigation, sailed over rivers and highroads and at last arrived at ye banks of ye Styx where we embarked on board Charon's boat and sailed 1,000

> yards underground by ye twinkling of a few farthing candles. We then arrived at a highroad leading to ye infernal regions where we met one of ye Furies of Darkness as black as ye Devil; by him we were conducted, having left our boats, 100 yards almost upon our hands and knees till we came to ye Coal . . .[24]

The idea of connecting the east and west coasts by canal was firmly in Stanhope's mind by the autumn of 1765, when he hired John Longbotham of Halifax (assistant to John Smeaton) to survey the lands on both sides of the Pennines. The Yorkshire side posed few problems, as Longbotham proposed a route from the end of the Aire & Calder Navigation at Leeds through Airedale to Gargrave. The Lancashire terrain proved more difficult, but in late December 1765 Longbotham promised Stanhope a completed preliminary survey by the following February. Several writers have suggested that the initiative for the canal came from the surveyor, who proposed the scheme to the energetic Quaker John Hustler. However, John Stanhope was committed to the canal before either of them, and several letters confirm his early interest and enthusiasm. Lowther at Preston described him as 'the great encourager of the intended Navigation', and Tempest of Tong Hall, Bradford, wrote to him, 'It is highly reasonable that the thanks of all whose property or dealings are interested in the intended navigation should be given to you as its author . . .'.[25]

Stanhope headed an early subscription list and addressed the merchants of Lancaster as early as the spring of 1766. He made several journeys to Liverpool that summer in order to influence the town's corporation and was asked by the committee to personally enlist the support of the Duke of Devonshire and the Earl of Thanet.[26] When the main subscription for the Leeds & Liverpool Canal was finally opened at Bradford in January 1769, John Stanhope headed the list with £4,000. Many of the West Yorkshire gentry and Bradford colliery owners contributed, but none put up more than a thousand pounds.[27] The Spencer-Stanhope family archive provides ample testimony to John Stanhope's close involvement and commitment to the scheme during the early promotional stage. Unfortunately he died in 1770.

The illness which ultimately caused his death forced Stanhope to relinquish his initiative in the canal as early as 1769. It was John Hustler who replaced him as the leading promoter. Hustler had become actively involved in the scheme in August 1766, having failed to attend the first official public meeting at the Sun Inn, Bradford, in July that year.[28] In attendance, however, were many of the principal landowners of Leeds and Bradford. Of those from Bradford no fewer than ten were working collieries on their estates at that time. The meeting approved Longbotham's plan and opened a subscription to cover initial costs.[29] Stanhope and Hustler were eager to obtain an Act of Parliament for the canal, for several other trans-Pennine schemes had been mooted. Respective county committees were formed, based at Liverpool and Bradford. The Yorkshire group included the Bradford colliery owners John Stanhope, Benjamin Ferrand and Henry Wickham. All parties spent the year 1767 mustering support for the Bill and, in particular, wooing the great landowners through whose land the canal was to pass.

Longbotham's survey (following the Ribble valley to near Preston and via Ormskirk to Liverpool) was sent to James Brindley for his criticisms and comments. In fact the work was done by Brindley's assistant, Robert Whitworth, in the summer months of 1768. Brindley's final report confirmed the route proposed by Longbotham, and it was accepted by the Yorkshire committee. A joint meeting was arranged for August that year, and Hustler, Balme and Buck were in attendance.[30] By this time the promoters of the canal in Lancashire were having serious doubts about the proposed route. As J. R. Harris has commented, 'The Yorkshire subscribers, for their part, wanted the speediest communication with Liverpool consistent with cheapness.'[31] Longbotham's plan had followed the Ribble valley and had not touched emerging industrial towns like Blackburn and Burnley (fig. 1). The Wigan coalfield could be penetrated only by the acquisition of the Douglas Navigation, and Lancashire industrialists were not to be fobbed off with branch canals.[32] As a result they quarrelled with the leading promoters, and this schism threatened the project. However, Hustler, Balme and William Thornton temporarily pacified the Lancashire committee, and on 19 December 1768 the two committees agreed to apply for an Act of Parliament. At this Burnley meeting the promoters accepted Brindley's estimated cost of £259,777 and opened a capital of £260,000 in £100 shares (with interest at 5 per cent). To open the subscription lists it was proposed to hold 'two great meetings', which took place simultaneously at Bradford and Liverpool on 9 January 1769.[33] Within a short time £150,000 had been subscribed, and it was decided to appoint Thomas Hardcastle, Thomas Leach and William Birkbeck as treasurers (all three were local bankers). Not long after these meetings Hustler informed Stanhope that people 'are generally desirous of better information', and he immediately drafted a pamphlet entitled 'A Summary View of the Proposed Canal'.

Throughout 1769 Hustler had the difficult task of keeping the two county factions in harness. The Lancastrians revived their complaint that the route went too far north and employed John Eyes and Richard Melling to plan a line that would meet their needs. Stanhope noted the growing dissent on all sides. He was informed that Hustler 'has lost his influence and others have taken up the lead which . . . will make confusion'.[34] Hustler asked Brindley to make a further report in the hope that it would placate dissidents. Again it was Whitworth who resurveyed the line, and he found errors in the survey completed by Eyes and Malling. The Bradford group then rejected the Liverpool alternative and insisted 'that the present association and subscription was entered into upon the plan produced at the meeting at Bradford and therefore this committee think they are not at liberty to depart from it . . .'.[35] Henceforth no other surveys were to be financed from the subscription. The Lancashire faction had meanwhile ordered another survey. It was awaiting P. P. Burdett's report when, at a general meeting summoned at the Sun Inn, Bradford, in October 1769, Hustler and Balme on behalf of the Yorkshire committee agreed that Brindley should consider both schemes. When Brindley confirmed Longbotham's

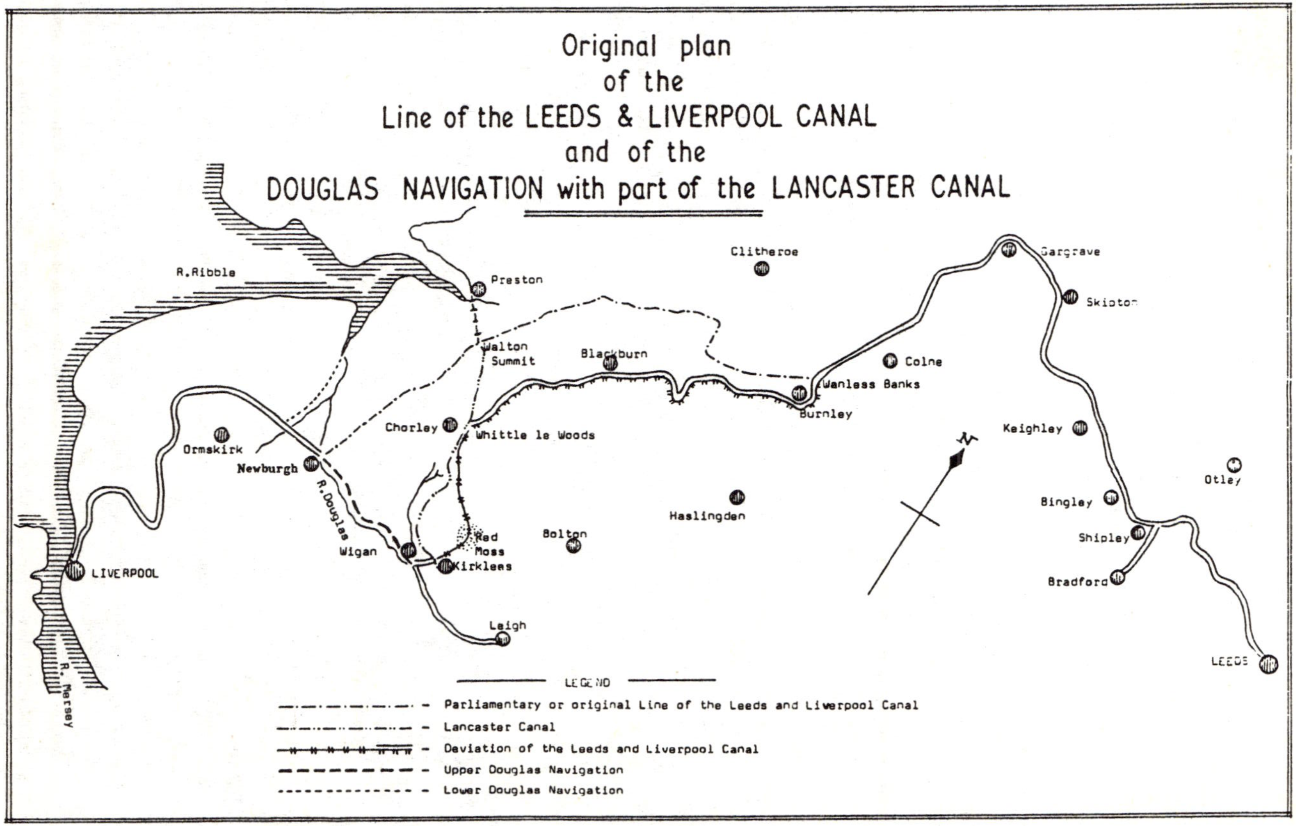

Fig. 1 Routes of the Leeds & Liverpool Canal, 1765–1816

route the Lancashire and Liverpool contingents withdrew their subscription, a decision confirmed at a meeting at Burnley on 11 December. Fearing further defections, Hustler chose to promote the Bill in London, and when a number of the Liverpool men surprisingly rejoined the committee in January 1770 Parliament was finally petitioned. The canal Bill became law in May 1770.[36] At the first general meeting a committee of twenty-three was elected, including Balme, Leach, Garforth, Walter Stanhope (John Stanhope was already dead) and Hustler, who was nominated treasurer. James Brindley was appointed chief engineer at £400 p.a., assisted by Longbotham as clerk of works at £160. Brindley was unable to accept the post (he died in 1772), and consequently Longbotham fulfilled both roles.[37]

Work began immediately on three sections which it was thought would provide the swiftest return on investment: Liverpool to the river Douglas, Skipton to Shipley and Shipley to Leeds. By early April 1774 the canal from Skipton to Thackley, near Shipley, was finished, together with the difficult stretch from Bingley to Shipley,

Plate 1 Spring's Canal. A branch canal, built by the Earl of Thanet in 1774, which went into the heart of the earl's limestone quarries at Skipton

Plate 2 Five Rise Locks, Bingley, completed in March 1774

and by 1777 the navigation was complete from Gargrave to Leeds. However, Brindley had seriously underestimated the cost, for by 1775 the company had already spent £232,000 in building only fifty-five miles of canal. It was forced to borrow £2,000 to pay contractors and meet the interest due to the proprietors. Hustler estimated that the remainder of the canal would cost £800,000, but he did not anticipate the economic recession that followed the colonial war in America. War debts needed funding, and consequently private credit was hard to come by as interest rates rose. For over a decade, from 1781, work stopped on the canal, which had been completed between Leeds and Skipton on the Yorkshire side and between Liverpool and Newburgh in Lancashire. The importance of this coalfield link to Liverpool has been stressed elsewhere. Coal from the Douglas valley 'became the lifeblood of this important and profitable waterway'.[38] However, the completion of the canal from Leeds to the Yorkshire dales was equally important for Bradford's economic development. While there was evidence of strong local support – after London, Bradford was the largest source of subscribed capital – Bradford colliery owners had been able to utilise a national subscription to promote and finance trading connections within their own regional economy. In particular, John Hustler had been able to exploit his connection with the Quaker communities in London and Norwich to Bradford's advantage.[39] Certainly, in the canal's early years, its affairs were run very much in the interests of Bradford. Joseph Priestley of Bradford was appointed accountant and clerk of works. The company's law clerk was John Eagles, who practised in Bradford, where the company's head office was located until 1850. John Hustler became sole treasurer of the company and controlled its finances until his death in 1790.

The influence of the Bradford colliery entrepreneurs spilled over the Pennines into the Lancashire coal industry. Several of the mid-Lancashire collieries opened in the decade after 1774 were leased by John Hustler, Thomas Hardcastle, John Jarratt, Joseph Dawson, John Hardy and Edmund Peckover. For one writer the Bradford-Liverpool caucus significantly transformed the capitalisation of Lancashire collieries:

> Local finance was dominant in the 1770s on the Leeds and Liverpool Canal when the industry grew slowly but large injections of capital from Liverpool and Bradford were associated with the more rapid growth of the last two decades of the century.[40]

John Hustler, who ran several collieries in Orrell and Up Holland as well as those around Bradford, dominated the affairs of the canal until his death. He and his clique of Bradford colliery owners were able to control the supply and distribution of coal at both ends of the canal as well as direct and manipulate company policy.

Although revenue came largely from the Lancashire side of the canal, the company's financial records for the late eighteenth century show a steady growth of income from the Yorkshire branch, a large share of which came from coal and limestone. The importance of that trade to the business of the canal can be assessed from the income on tolls. In 1792 coal and limestone accounted for 40·7 per cent of the gross income from tolls along the whole of the canal. By 1820, when the canal was completed, it had increased to 50·2 per cent, and by the 1830s remained at least 40 per cent of the income. Trade of this kind was realised by the Bradford

Table 1 *Leeds & Liverpool Canal Company accounts, 1772–1800 (quinquennial averages)*

Period	*Total gross income (£)*	*Income from Yorkshire side (£)*	*% total*
1772–75	1,633	932*	–
1776–80	6,393	2,561	40
1781–85	9,746	2,745	28
1786–90	16,892	3,943	23
1791–95	25,160	5,350	21
1796–1800	36,298	8,813	24

*1773–75

Source. 'Statistical History of the Leeds & Liverpool Canal' (1898), in Rail 846/124, P.R.O.

colliery owners only after the town had been linked direct to the Leeds & Liverpool Canal. There can be little doubt that this was their intention all along. Early in 1770 fourteen Bradford merchants and landowners met to promote a Bill for a Bradford canal; six (including Hustler and Balme) were already members of the Leeds & Liverpool committee.[41] In April 1771 an Act was passed to connect Bradford by water to the main canal. This was achieved between 1771 and 1774 with the construction of the small branch between Windhill and Bradford.[42] James Brindley visited Bradford in 1770 and proposed a three-mile route which was immediately approved. Of the original subscribers the majority were colliery owners and included the familiar entrepreneurial names of Leach, Balme, Leedes, Preston and John Hustler, who once more played a prominent part in raising the requisite £6,000 for the completion of the canal.[43]

The creation of the Bradford Navigation Company led to the founding of the Bradford Limekiln Company in March 1774. There were ninety shares in the lime-burning concern, of which more than two-thirds were owned by John Preston, Abraham Balme, William Hardcastle, John Hustler and Edward Leedes, all coal traders and substantial shareholders in both the local canals.[44] The amount of limestone purchased and consumed by the company increased from 3,858 tons in 1774 to 14,686 tons two years later. The importance of the new canal network to the local lime industry was noted in the press:

> On the Yorkshire side we found eighteen boats of burthen already built, a number of which have been for some time fully employed between Skipton and Bingley in the lime and coal trade. There are forty lime-kilns now erected and preparing between Skipton and Bradford and the business on the line in this article and in coal must be very large.[45]

Until the crisis of 1781 the Bradford Limekiln Company achieved a good trade, increasing its sales of lime from 4,621 tons in 1774 to 53,257 tons in 1778.[46] Thereafter the company ran into difficulties, probably of its own making, since the proprietors were trading as individual customers and suppliers. By 1785 the affairs of the Bradford canal and the limekiln company had been taken over by the joint-stock company of Preston Hird & Jarratt, a concern which had operated since May 1778 to develop some of the mineral resources at Low Moor. When Edward Leedes's manorial estate at Low Moor came on to the market this syndicate was an appropriate candidate for ownership, and was subsequently, as the Low Moor Ironworks Company, the first successful large-scale industrial enterprise in Bradford.[47]

Their joint-stock trading was the key to future success, but before 1778 each of them had operated individually in the local coal–limestone trade. The success of the Low Moor Ironworks, and of others like it in the Bradford region after 1790, stemmed originally from the individual activities of a small number of landowners anxious to play to the full the role of industrial entrepreneur. In order to break out of the straitjacket of the local transport network they involved themselves after 1740 in the promotion of turnpike trusts, banks and, most important of all, in the development of a regional canal system. Their efforts richly enhanced the efficiency of the regional economy as a whole, and of Bradford in particular. The West Riding of Yorkshire became one of the most important growth regions of industrial Britain in the nineteenth century. The economic activities of a small group of Bradford coal owners in the 1760s went some way in preparing the seedbed of that growth.

Notes

1 J. H. Farrington, 'The Leeds and Liverpool Canal: a study in route selection', *Transport History*, III (1970), 53. Other studies which have referred to Bradford's connections with the canal include J. R. Harris, 'Liverpool canal controversies, 1769–1772', *Journal of Transport History*, II (1956), 158–72. See also C. Hadfield and G. Biddle, *The Canals of North West England*, I (1970), p. 65; L. T. C. Rolt, *Navigable Waterways* (1973), pp. 56–9; R. W. Unwin, 'Leeds becomes a transport centre', in D. Fraser (ed.), *A History of Modern Leeds* (1980), pp. 113–14.

2 J. Priestley, *A Historical Account of the Navigable Rivers, Canals, etc.* (1831), repr. (1969), pp. 385–97.

3 H. F. Killick, 'Notes on the early history of the Leeds and Liverpool Canal', *Bradford Antiquary*, new series, III (1897), 169–238.

4 J. James, *A History and Topography of Bradford* (1841), p. 1.

5 C. Richardson, *A Geography of Bradford* (Bradford, 1976), p. 47.

6 W. Cudworth, *Round about Bradford* (Bradford, 1886), pp. 56–7, and A. L. Green, *Geology of the Yorkshire Coalfield* (1878), pp. 658–76.

7 B. G. Auty, 'Sir Richard Shireburn's Esholt Ironworks', *Bradford Antiquary*, new series, VIII (1960), 243–57.

8 G. Firth, 'The Genesis of the Industrial Revolution in Bradford', unpublished Ph.D. thesis, University of Bradford, 1974, pp. 237–67.

9 York Minster Library (hereafter Y.M.L.), Hailstone Mss, BB 42.1.

10 Y.M.L., Hailstone Mss, HH 5.31 and PP 5.

11 Y.M.L., Hailstone Mss, BB 42.1.

12 W. Cudworth, undated note in Bradford Central Library (hereafter B.C.L.)., Miscellaneous Mss, Indenture, 12 January 1775.

13 A. Raistrick, *Old Yorkshire Dales* (1967), pp. 67–81.

14 W. E. Preston, 'Some notes on an old Bradford firm', *Bradford Antiquary*, new series, V (1907–11), 312.

15 T. C. Barker and C. I. Savage, *An Economic History of Transport in Britain* (1974), pp. 20–3.

16 Parliamentary Papers, Journal of the House of Commons, 1 February 1744.

17 Richardson, *op. cit.*, pp. 48–51.

18 J. Hustler, *A Summary View of the Proposed Canal from Leeds to Liverpool* (Leeds, 1770), p. 10.

19 B.C.L., Scruton Collection, 16/1/14.

20 *York Courant*, 7 August 1764.

21 T. C. Barker, 'The beginnings of the Canal Age in the British Isles', in L. S. Pressnell (ed.), *Studies in the Industrial Revolution* (1960), pp. 1–22.

22 C. Hadfield and G. Biddle, *The Canals of North West England* (Newton Abbot, 1970).

23 R. G. Wilson, 'Three brothers: a Study of the fortunes of a landed family in the mid-eighteenth century', *Bradford Textile Society Journal* (1964–65), 111–22.

24 B.C.L., Spencer-Stanhope Ms 2246/65, Walter Stanhope to John Stanhope, 2 July 1765.

25 *Ibid.*, 1170, undated letter, John Longbotham to John Stanhope.

26 *Ibid.*, 2246, various letters, John Tempest to John Stanhope.

27 Minute Book of Leeds Liverpool Canal Company, 9 January 1769, Rail 846/2, Public Record Office (P.R.O.).

28 *Ibid.*, 14 July 1766.

29 *Leeds Intelligencer*, 5 August 1766.

30 C. T. G. Boucher, *James Brindley, Engineer, 1716–72* (1968), p. 68.

31 Harris, *loc. cit.*, 161.

32 J. Langton, *Geographical Change and Industrial Revolution* (Cambridge, 1979), pp. 169–74.

33 Leeds & Liverpool Canal, Letter Book, 19 December 1768, Rail 846/2, P.R.O.

34 B.C.L., Spencer-Stanhope Ms 1170, John Hustler to John Stanhope, 24 December 1768.

35 Leeds & Liverpool Canal Minutes, 24 July 1769, Rail 846/2, P.R.O.

36 10 Geo. III, cap. 114; Hadfield and Biddle *op. cit.*, pp. 70–1.

37 Leeds & Liverpool Canal Minutes, 19 July and 31 August 1770, Rail 846/2, P.R.O.

38 Langton, *op. cit.*, p. 170.

39 Alphabetical List of the Proprietors of the Leeds-Liverpool Canal, 1770, 1789, 1800, B.C.L., Misc. Ms. See also E. M. Sigsworth, 'Bradford on the eve of the industrial revolution', *Bradford Textile Society Journal* (1953–54), 61–4; A. Raistrick, *Quakers in Science and Industry* (Newton Abbot, 1968), p. 78.

40 Langton, *op. cit.*, p. 224.

41 Bradford Canal Company Minute Book, 1770–1, B.C.L.

42 W. Robertshaw, 'Historical links between

Bradford and Shipley', *Bradford Antiquary*, new series, IX (1950), 277–92, and J. R. Roberts, 'Bradford waterways: rise and fall', *Bradford Textile Society Journal* (1962–63), 147–55.

43 Bradford Canal Company Minute Book, 9 August 1771, B.C.L., and 2 Geo. III, cap. 89.

44 See G. Firth, 'The Bradford Limekiln Company, 1774–1800: a pioneer of large-scale industrial enterprise in Bradford', *Bradford Antiquary*, new series, XLVII (1982).

45 *Leeds Intelligencer*, 28 June 1774.

46 Bradford Limekiln Company Papers, Accounts Ledger, 1774–97, B.C.L.

47 G. Firth, 'The origins of the Low Moor Ironworks, Bradford, 1788–1800', *Yorkshire Archaeological Journal*, XLXIX (1977), 127–39.

Acknowledgements

The author is grateful to the archivists and staff of the Public Record Office, the York Minster Library and Bradford Central Library (Archives) for their assistance during the course of research, and to Ian Dick for mapwork.

5 *Transport change and the development of the Leicestershire coalfield in the canal age: a reinterpretation*

C. P. GRIFFIN

THERE is an extensive literature on the economic history of the Leicestershire coalfield in the so-called canal age.[1] There is general agreement on the broad outline of this history and the central propositions may be summarized briefly as follows. Before the development of canals in the east Midlands the Leicestershire coalfield was able to compete effectively and had a virtual monopoly of the county market, but with the coming of canals the coalfield at worst ceased production[2] and at best totally stagnated[3] because the canal built to serve it was technically inadequate, and it was therefore unable to take advantage of an improved transport network. This dismal state of affairs endured until either, according to some sources,[4] the coalfield was revived from the mid-1820s by the prospect of using railway transport, or according to others remained closed until the railways were actually opened in the early 1830s.[5] The purpose of the present contribution to the already voluminous literature concerned with the relationship between transport improvement and the market for Leicestershire coal is to suggest that the conventional interpretation is at several points in need of revision.

I

In the mid-eighteenth century the market for Leicestershire coal was a regional one protected by high transport costs from coal produced by the coalfields of east Warwickshire and south Derbyshire, situated to the south and west, and the Erewash Valley coalfield to the north. The Leicestershire coal output of between 30,000 and 50,000 tons dominated the county market and penetrated into Lincolnshire and Northamptonshire, though high transport costs still meant that the inhabitants of the remoter areas of the county continued to burn poor substitutes such as wood and cow dung.[6] From the late 1770s the market situation changed rapidly with the opening of the Erewash canal (1779) and the Soar Navigation (1778), which enabled waterborne coal from the Erewash Valley to enter the Leicestershire market through Loughborough, in particular, at a much reduced price (£1 to 13*s*. 4*d*. (66½p) a ton).[7] By 1785 it was claimed that the 60,000 tons of coal used each year in Leicestershire, Rutland and parts of Northamptonshire were supplied by the Leicestershire and Erewash Valley coalfields in roughly equal proportions;[8] not surprisingly the

Leicestershire owners were very worried indeed over the various proposals that were being made to extend the Soar Navigation from Loughborough to Leicester itself in order, it was claimed, to reduce the cost of carriage of coal between the two places from 5*s.* to 1*s.* 6*d.* (25p to 7½p) a ton; they could see their market being swamped by coal coming in from the north.[9] The Leicestershire colliery owners strenuously opposed several proposals which were made between 1785 and 1789 to construct a navigation between the two towns because they could not safeguard their interests, and it was not until 1790 that they were reconciled with other groups including several large landowners such as Lord Rawdon and Earl Ferrers, leading Leicester businessmen and bankers, and the Soar Navigation Company, and agreed to give their support to the formation of the Leicester Navigation Company. The Company was empowered to build two connecting levels, one from Loughborough to Leicester and another from the Leicestershire coalfield at Thringstone to Loughborough – the so-called Charnwood or Forest line – though because of geological difficulties the last stretch of this section of the whole navigation from Nanpanton to Loughborough would be a railway. Neither canal was allowed to carry coal until both were complete and this device would ensure that both coalfields were on an equal competitive footing in the Leicestershire market.[10]

In October 1794 both navigations were declared open though the Forest line was, in fact, far from finished and in particular suffered from a shortage of water since its main reservoir at Blackbrook had not been completed. As a result the Leicestershire owners had considerable difficulty selling their coal at a competitive price at Loughborough, and they eventually received assistance from the Leicester Navigation Company who went into the coal merchanting business on their behalf and also gave a general reduction in tolls in mid-1797 because adequate coal traffic was still not forthcoming.[11] In February 1799 efforts to encourage traffic were dashed when the reservoir burst at the head and the released water destroyed an aqueduct and a stretch of the canal as well as submerging land for miles around.[12] The reservoir head was rebuilt but the canal remained unused; after the reservoir was emptied in 1804 in order to prevent a possible repetition of the flood, [13] it fell into decay.

The failure of the Forest Line has been a subject of speculation among historians ever since. Two nineteenth-century historians of the Midland railway, C. E. Stretton and F. S. Williams, concluded in sweeping style that the canal burst its banks and was never repaired with the result that 'the failure of the branch canal effectively shut out Leicestershire coal and gave the entire trade to the Nottinghamshire and Derbyshire coal owners, a condition of things that remained unaltered for no less than thirty three years'.[14] This view has been endorsed by other historians in more recent times,[15] though others have observed that the canal was restored following the flood but that it was not a success because the colliery owners lacked the confidence that their coal could be sold down it at a profit, since its use as far as Loughborough involved a three-fold[16] transhipment of the canal, compared with the Erewash Valley's one. Because the Forest Line was a technical failure, the Leicestershire owners were, in effect, deprived of a sufficiently cheap and efficient transport network with the result that there set in a period of 'almost complete paralysis for over thirty years from the lack of cheap transport'.[17]

This interpretation of events which stresses the technical failure of the Forest Line as the root cause of the coalfield entering a period of paralysis rather surprisingly ignores the view of that generally well-informed contemporary John Farey, Jr, who noted in 1811 after a visit to the area that the failure of the Forest Line was not brought about by any deficiencies in its construction, which then caused the colliery owners to lose faith in the idea that their coal could compete with that of their rivals, but was a direct consequence of the closure of the chief collieries in the course of a lawsuit which deprived the canal of traffic.[18] Farey's observation suggests that it was the economic and social circumstances of the colliery owners that provides the key to an understanding of what actually occurred and it is to this line of enquiry that we must now turn our attention.

II

The principal collieries for whose benefit the Forest Line was constructed were those at Coleorton and Newbold, leased by Joseph Boultbee from Sir George Beaumont; Swannington, leased by Messrs. Fenton and Raper from the trustees of Wigston Hospital, Leicester; and Thringstone, leased by Godolphin Burslem from Beaumont.[19] In 1792 William Black was employed by Beaumont to make a thorough survey of his estate and investigate the affairs of his steward Joseph Boultbee, and Black's report resulted in Beaumont accusing Boultbee, and before him his father, who had been steward for 32 years prior to his death in 1790, of fraudulent accounting, concealing information so as to obtain ridiculously low leasehold terms for the collieries, and selling timber and clay from the estate for their own profit; Beaumont claimed £20,000 damages, although he considered that the Boultbees had swindled his family out of a much larger sum during their years of employment with them. Boultbee argued that if he and his father had received a fairer salary than the £20 a year they had been paid since 1758 then they would not have been dishonest, and he made a determined attempt to continue to work the collieries following his dismissal in late 1792, although he was on rather difficult ground since the leases had been renewed in 1781 on 'an informal basis'. Beaumont managed to gain control of Newbold in a matter of months only to find that the existing seams were exhausted, but he was prevented from seizing Coleorton by Boultbee's threat to flood the workings. When Boultbee was forced to relinquish the colliery in 1797 he dismantled the large pumping engine which had been installed in 1781 at a cost of £1,829 and rendered the colliery unworkable without heavy capital expenditure.[20] Beaumont's annual income was only about £8,000 and he tried to find a lessee for Coleorton since he did not feel that his own financial resources were sufficient to get the colliery working again. A lessee was not forthcoming and Beaumont only received about one-third of the damages which he had claimed from Boultbee, with the result that the colliery remained closed until 1820.[21]

The mid-1790s was also a critical period for Burslem. In 1795 his enterprise was being hampered by the failure of the Forest Line to function efficiently, but this was not too serious a problem since the Leicester Navigation Company were buying a guaranteed weekly output from his colliery at market prices until the reservoir was built.[22] None the less, Thringstone colliery had closed by 1796 and it is generally assumed that Burslem had 'gone out of business' because of a failure to sell coal down the Forest Line at competitive prices following the completion of the reservoir.[23] This assumption is highly questionable, however, since Burslem was working 'a fine new mine' at Pegg's Green in 1805 with the assistance of 'an exceedingly good new engine which throws up 760 gallons of water a minute'.[24] It seems more likely, in fact, that rather than going out of business in 1796 as a result of marketing problems which were probably as acute in 1805 as a decade earlier, Burslem's colliery had been developed as far as the massive Thringstone uplift fault – at which all coal seams disappeared and whose existence hitherto had not been realized – and that when faced with this insuperable natural obstacle he simply cut his losses and sank a new colliery about 1½ miles away from the fault at Pegg's Green.[25]

The lessees of Swannington colliery, Fenton and Raper, were not local capitalists like their neighbouring colliery owners; both originated from the West Riding and as well as being partners in various enterprises in the Leeds area – including Middleton colliery – they also owned a colliery and ironworks at Aldercar, Derbyshire.[26] Their commitments were both extensive and geographically widespread and it is likely, indeed, that they had over-extended themselves and attempted to rectify this position in the 1790s. In September 1793 they sold their Aldercar concern and no doubt regretted their investment of £400 in the Leicester Navigation Company, since it had not yielded the worthwhile dividend which had been predicted during its planning stage.[27] From the mid-1790s Swannington colliery was approaching exhaustion and in July 1799 the remainder of its lease was offered for sale. They proposed to the Leicester Navigation Company that they should take it over 'if they should think it proper to begin a new foundation on their own account'.[28] No buyers were forthcoming and so the colliery was closed, the partners were not themselves prepared to invest further capital in it, and in March 1800 they withdrew from the county altogether after making a final sale of all the 'cattle, sheep and implements' belonging to their Swannington estate.[29]

The widely-held belief that the failure of the Forest Line can be explained by the adverse marketing conditions it produced for the colliery owners is open to serious doubt. Rather it seems that it was the difficulties which the collieries in the vicinity of the canal were facing for reasons unconnected with transportation which guaranteed that it was so little used that Blackbrook reservoir was dismantled only five years after its restoration, after which the canal fell into complete disrepair. It is true that Leicestershire coal had difficulty in competing in the Leicester market because its more frequent transhipment than Erewash Valley coal meant higher transport costs and poorer quality,[30] but once the Forest Line was fully operational it seems likely that this problem would not have proved an insuperable one if only the collieries for which it had been constructed had remained in an efficient working condition. After all, the pithead price of Erewash Valley coal was continually advancing[31] and would soon have given the Leicestershire owners a price advantage, and the Leicester Navigation Company, too, were willing to adjust the level of tolls on their navigation as a whole in order to bring Leicestershire coal into competition with that of its rival at a profitable price.[32] Moreover, in May 1798 Beaumont was negotiating favourable terms with the Leicester Navigation Company in anticipation of the successful outcome of his lawsuit with Boultbee[33] and there seems little reason to doubt that Coleorton colliery would have been reopened and sold its output down the canal had the lawsuit gone more in Beaumont's favour.[34]

III

The failure of the Forest Line is generally seen as a signal for the complete, or at least partial, closure of the coalfield for almost 30 years. This view seems to stem from the fact that the collieries for which the navigation was built remained closed until the coming of railways was anticipated in the mid-1820s – or even until they were actually built in the early 1830s, according to some accounts. Now it is perfectly true that Coleorton was not re-opened until the 1820s, Swannington until the early 1830s, and that Thringstone ceased to be a mining centre at all, but the weakness of this approach is that it focuses too narrowly on the original colliery complex to the exclusion of conditions prevailing elsewhere in the coalfield. The collieries of Earl Ferrers at Lount and Staunton Harold, situated about two miles north-west of the Coleorton/Swannington complex, continued working throughout the whole period greatly assisted by their connection to the Ashby canal.[35] In 1802 these mines were said to be 'working to considerable effect' under the stimulus of a price rise in the last few years of from 6*s*. to 20*s*. (30p to £1) a load,[36] while there was an important new mine working at Pegg's Green from 1805.[37] There were smaller mines, too, at Heather and Hugglescote which supplied a local market because their output was 'not deemed to be of the first quality'.[38] In 1820 Beaumont decided to risk his own capital in a re-opened Coleorton mine presumably because he was still unable to find suitable lessees following the Boultbee litigation. Improved pumping machinery was installed with the assistance of miners brought down from Northumberland and coal was turned at a depth of 100 yards.[39] Sir Charles Abney likewise expanded his activities at Heather in 1820 with the sinking of a deeper mine to better-quality coal,[40] and shortly afterwards a partnership between John Whetstone, a Leicester merchant, and John Stenson, mining engineer, was formed to sink a colliery at Whitwick which opened in 1824.[41] The following year an Ibstock landowner, John Thirlby, opened a large colliery on his estate,[42] and in 1826 Lord Maynard followed suit at Bagworth.[43] Meanwhile the partnership of

Price, Bostock and Collishaw started improving and extending their newly acquired Pegg's Green Colliery and by 1830 five new seams had been opened out.[44]

Clearly the Leicestershire coalfield experienced a sharp setback with the closure of Colcorton, Thringstone and Swannington collieries between 1796 and 1800, but equally clearly the coalfield was enjoying really substantial growth in the 1820s and never experienced the desperate situation conjured up for it by the most recent historian of the Forest Line, who writes of the situation in 1799: 'the coal owners had had enough. With little or no profit realised over the past few years, they closed down their pits entirely, leaving the Derbyshire coal owners with a complete monopoly until the coming of the railways nearly thirty years later.'[45] On the contrary, all the pits were never closed and new ones were opening up from at least 1805.

Underlying the view that the coalfield ceased production is the idea that carting by road made the price of Leicestershire coal too high to compete with that brought to market by canal from the Erewash Valley,[46] but this assumption has never been seriously examined. In fact the experience of the turnpike roads which served the coalfield, and which were heavily dependent upon coal traffic for their revenue, suggests that this premise is in need of modification, if not actual rejection. It is a reasonable expectation that if the Leicestershire coalfield had ceased production over a long period it would have severely depressed the income of the turnpikes and the amount for which the toll bars were let annually.[47] In fact a sustained depression in turnpike revenue did not occur, nor were there any difficulties in letting toll bars.

The coalfield was served by three major turnpikes – the Ashby to Leicester, the Ashby to Loughborough and the Hinckley to Melbourne Common[48] – and on the whole their income and toll bar lettings follow the pattern we should expect, given the state of the Leicestershire coal industry outlined above.

TABLE I

Income and Toll Lettings of Turnpike Roads Serving the Leicestershire Coalfield 1789–1834

Date	Hinckley to Melbourne Common Toll lettings Swannington area bars	Loughborough to Ashby Toll lettings Thringstone	Loughborough to Ashby Toll lettings Coleorton	Leicester to Ashby Total Income
	(£)	(£)	(£)	(£)
1789	477			392
1790	479			—
91	—			—
92	—			—
93	570			—
94	449			669
95	—			769
96	550			492
97	—	155	191	419
98	—	156	165	—
99	302	172	183	—
1800	265	133	122	—
01	267	135	140	—

02	293	—	—	520
03	295	171	95	—
04	221	—	—	580
05	336	118	116	476
06	386	—	—	463
07	366	—	—	505
08	390	155	98	672
09	386	132	103	669
1810	386	126	113	626
11	503	—	—	642
12	386	151	107	656
13	370	—	—	608
14	385	—	—	709
15	—	—	—	—
16	—	—	—	808
17	—	—	—	853
18	—	—	—	923
19	—	—	—	—
1820	—	159	132	913
21	—	131	133	—
22	—	267	116	834
23	337	170	163	—
24	310	—	—	1,028
25	395	—	—	994
26	395	172	192	—
27	—	—	—	1,051
28	—	—	—	1,033
29	—	—	—	1,171
1830	—	—	—	1,132
31	—	—	—	1,306
32	—	—	—	1,315
1834				1,179

— Not known.
Source: *Leicester Journal* (toll lettings), Minute Book of Leicester to Ashby Turnpike Road.

The income of the Ashby to Leicester turnpike gradually increased from the late 1780s until the Leicester Navigation was opened in 1795. The drop in income experienced in 1798 was partly due to the fact that the letting of one of the chief gates, that at Markfield, was auctioned for £110 less than in the previous year 'in consequence of the opening of the navigation from Loughborough to Leicester'.[49] Income had partly recovered by 1802 and a new toll bar was erected at Whitwick in 1806 on account of increased coal traffic.[50] By 1808 income was back at about its pre-coal level and continued to rise thereafter until the opening of the Leicester to Swannington railway in 1833.

The amount for which the toll bars at Thringstone and Coleorton, situated on the Ashby to Loughborough turnpike, were let suggests that coal traffic continued to pass along the road although in diminished quantities at the turn of the century. Thereafter, although the lettings of both gates fluctuated considerably, they remained fairly buoyant and it is significant that the letting of the Thringstone gate fared better than that at Coleorton – probably because it was nearer to Pegg's Green and Lount collieries – and that the amount for which the Coleorton gate was let increased after Coleorton colliery was re-opened.

An indication of the amount of coal carried on the Melbourne Common to Hinckley turnpike is provided by the lettings of the two toll bars in the Swannington area, and here the lettings were buoyant until 1796 but had greatly declined by 1799 and did not show a recovery until 1805, the year Pegg's Green colliery was opened. The lettings continued to be relatively buoyant thereafter although at a lower level than they had been in the early 1790s.

IV

The relationship between transport development and the economic state of the coalmining industry in Leicestershire is clearly more complex than the historians of industry and transport in the county have so far supposed. It can no longer be uncritically accepted that the failure of the Forest Line was due to its own design features and that its failure, in turn, destroyed the economic viability of the Leicestershire colliery owners. It may well be the case too that, although when Sir Robert Southall read a paper before the Royal Society in London in 1675 the comparative costs of carrying heavy freight were such that you could carry a cargo of coal for 300 miles by water for the same costs which would have taken it only 15 miles by road,[51] it was not, apparently, physically impossible for Leicestershire coal using road transport to compete with coal from neighbouring counties using waterways at the beginning of the nineteenth century, as historians have hitherto supposed. The number of carters' horses ill-treated or even flogged to death on the Ashby to Leicester turnpike is testimony to the great efforts which were required to maintain this rather unequal competition in the closing years of long-distance horse-powered transport systems.[52]

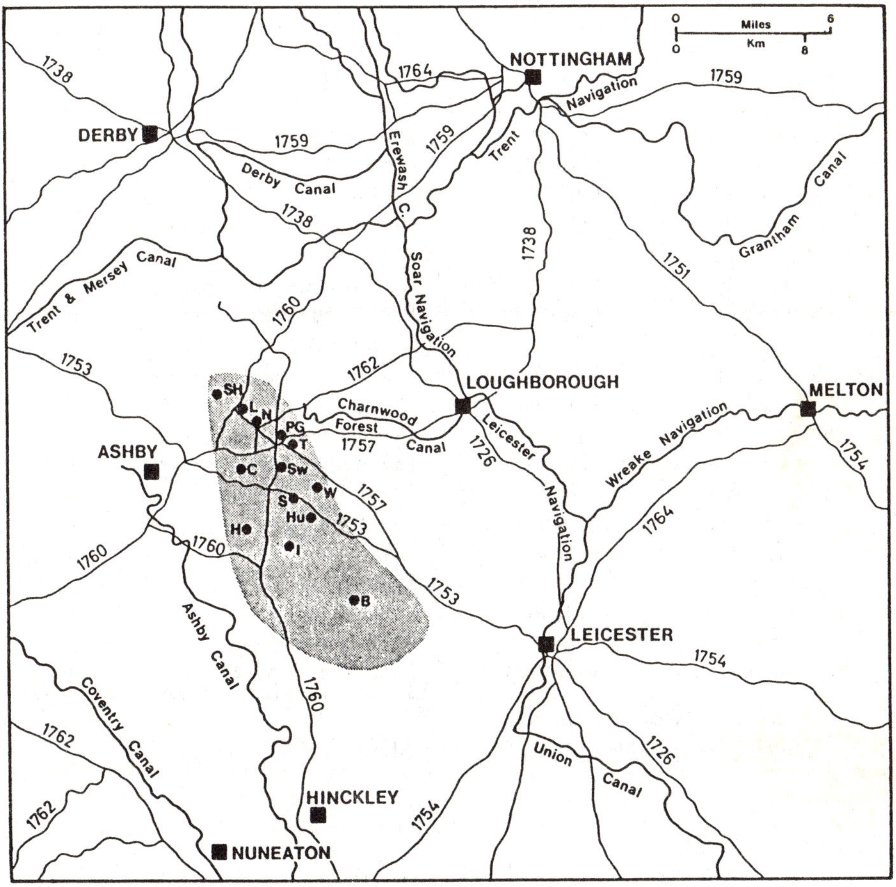

Coalfield — 1760 Turnpikes — Canals

COLLIERIES

SH	Staunton Harold	L	Lount	Sw	Swannington
N	Newbold	PG	Peggs Green	Hu	Hugglescote
T	Thringston	C	Coleorton	H	Heather
W	Whitwick	S	Snibston	I	Ibstock
B	Bagworth				

The transport system around the Leicestershire coalfield, c. 1800.

APPENDIX. PRINCIPAL SHAREHOLDERS OF THE LEICESTER NAVIGATION COMPANY IN 1791*

Name	Address	Occupation	Number of shares
Lord Rawdon	Donington Park	Landowner, colliery owner	20
R. Bree	Leicester	Doctor of physic	12
J. Carter	Leicester	Gentleman	11
Mary Wynne	Gumley	Widow	11
Lord Stamford	Enville	Landowner, lime works owner	10
Earl Ferrers	Staunton Harold	Landowner, colliery owner	10
Joseph Boultbee	Leicester	Banker, colliery owner, land agent	10
Duke of Newcastle	Nottingham	Landowner, colliery owner	10
J. Burbidge	Leicester	Hosier	7
S. Clerk	Leicester	Gentleman	6
Sir George Beaumont	Colcorton	Landowner, colliery owner	5
T. Babington	Leicester	Hosier	5
G. Burslem	Ravenstone	Colliery owner	5
F. Beresford	Derby	Gentleman	5
W. Herrick	Beaumanor	Landowner	5
T. Lockwood	Leicester	Gentleman	5
T. Pares	Leicester	Banker, landowner	5
Joseph Wilkes	Measham	Landowner, colliery owner	5

* Holding five or more shares.

Source: Leicester Navigation Company Minute Book, July 1791.

NOTES

1. The chief contributions are as follows: T. J. Chandler, 'Communications and a coalfield', *Transactions of the Institute of British Geographers* (1957); T. J. Chandler, 'The canals of Leicestershire. Their development and trade', *East Midland Geographer* (1958); F. G. Farnham, *Charnwood Forest and Its Historians* (1958); C. Hadfield, *The Canals of the East Midlands* (1966); W. D. Holmes, 'The Leicestershire and South Derbyshire Coalfield. (1) The coal mining industry', *East Midland Geographer* (1958); T. R. Potter, *Charnwood Forest* (1842); C. E. Stretton, *A History of the Midland Railway* (1901); A. Temple Patterson, 'The making of the Leicestershire canals 1766–1814', *Transactions of Leicestershire Archaeological and Historical Society* (1951); C. J. Williams, *The Forest Line. An Exploration of the Lost Charnwood Forest Canal* (1974); F. S. Williams, *The Midland Railway* (1878).
2. C. J. Williams, *op. cit.*, 12, for example.
3. Chandler, 'Communications and a coalfield', 168, for example.
4. Temple Patterson, *op. cit.*, 90, for example.
5. F. S. Williams, *op. cit.*, 6, for example.

6. *Leicester Journal*, 10 and 24 December 1785; W. G. Hoskins, *The Midland Peasant* (1965), 192.
7. Temple Patterson, *op. cit.*, 68–9.
8. *Leicester Journal*, 10 December 1785.
9. Temple Patterson, *op. cit.*, 70.
10. *Ibid.*, 73–5. The principal shareholders in the Leicester Navigation Company reflect those groups which supported its flotation, and relevant information about them is given in the Appendix.
11. Leicester Navigation Company Minute Book 5 January 1795, 3 July 1797.
12. *Ibid.*, 22 February 1799.
13. Temple Patterson, *op. cit.*, 90.
14. Stretton, *op. cit.*, 4; F. S. Williams, *op. cit.*, 6.
15. For example in Holmes, *op. cit.*, 21.
16. Temple Patterson, *op. cit.*, 86–7.
17. Chandler, 'Communications and a coalfield', 170.
18. J. Farey, *A General View of the Agriculture and Minerals of Derbyshire* (1811–17), III, 38.
19. C. P. Griffin, 'The Economic and Social Development of the Leicestershire and South Derbyshire Coalfield 1550–1914 (unpublished Ph.D. thesis, University of Nottingham, 1969), 202–9.
20. P.R.O., C 12/255/1, 1797–1800.
21. E. M. Howe, 'Coal, art and the Beaumonts', *History Today*, XXIV (1974), 247.
22. Leicester Navigation Co. M.B.5, January 1795.
23. Temple Patterson, *op. cit.*, 87.
24. J. Nichols, *The History and Antiquities of the County of Leicester* (1811), 740.
25. C. Fox-Strangeways, *Memoirs of the Geological Survey of England and Wales. The Leicestershire and South Derbyshire Coalfield* (1907), 21.
26. G. Rimmer, 'Middleton Colliery, near Leeds, 1770–1830' *Yorkshire Bulletin of Economic and Social Research* (1956), 40, *Victoria County History, Yorkshire*, II (1960), 261, 330, 357 and 400.
27. *Leicester Journal*, 6 September 1793; Temple Patterson, *op. cit.*, 86.
28. Leicester Navigation M.B., 12 July 1799.
29. *Leicester Journal*, 12 March 1800.
30. Leicester Navigation Company Letter Book, Carter to Deakin, 6 March and 15 June 1796. The Blackbrook reservoir was completed in June 1796.
31. R. Lowe, *A General View of the Agriculture of Nottinghamshire* (1798), 134.
32. Leicester Navigation M.B., 3 July 1797.
33. *Ibid.*, May 1798. The Company in fact acceded to Beaumont's terms which are outlined in Griffin, *op. cit.*, 856.
34. Beaumont claimed £20,000 damages and the return of Coleorton colliery in a workable condition, but in fact received only about £6,000 damages and a colliery which was drowned out and totally devoid of a pumping engine and other equipment. P.R.O. C 12/255/1 1797–1800, Howe, *op. cit.*, 247.
35. Reprint of the First Edition of the One-Inch Ordnance Survey of England and Wales, sheet number 43 (1968).
36. Nichols, *op. cit.*, 718.
37. *Ibid.*, 740.
38. *Ibid.*, 931; J. Throsby, *Leicestershire Excursions* (1790), 478.
39. A. Temple Patterson, *Radical Leicester* (1954), 260.
40. F. White, *History, Gazateer and Directory of the County of Leicester* (1846 edn), 549.
41. Rev. J. Curtis, *A Topographical History of the County of Leicester* (1831), 184.
42. White, *op. cit.*, 564.
43. *Ibid.*, 596.
44. *Ibid.*, 369.
45. C. J. Williams, *op. cit.*, 12.

46. Chandler, 'Communications and a coalfield', 168, for example.

47. For a similar attempt to calculate the volume of traffic using the main turnpike roads around Leeds between about 1770 and 1820 on the basis of toll lettings, see R. G. Wilson, 'Transport dues as indices of economic growth 1775–1820', *Economic History Review* (1966), 110–23. The argument is based upon the following propositions: 1. Coal was almost certainly the most important single traffic on the Loughborough to Ashby, Ashby to Leicester and on the coalfield section of the Hinckley to Melbourne Common turnpike roads. Some indication of its importance can be derived from the fact that it was reported in July 1808 that the income of the Groby and Whitwick Long Lane toll gates on the Ashby to Leicester turnpike fell off sharply 'in consequence of the discontinuance of the working of the coal pits', and when Sir George Beaumont constructed a new private road from Colcorton colliery to the Loughborough to Ashby turnpike in 1822 and closed the one to the Hinckley to Melbourne Common turnpike several repair men on the latter were made redundant because of the loss of traffic. 2. Toll bars were let annually to the highest bidder and were the turnpikes' principal source of revenue. The fluctuations in the amount for which they were let reflected the level of income which the lessee (or farmer, as he was called) expected the bar to yield, and since toll charges were fixed by Act of Parliament and were not generally increased during the period, the amounts for which the gates were let would change according to the level of expected economic activity and hence traffic in the area, and, as will be argued below, on the output of the coal industry in particular: Ashby to Leicester turnpike Minute Book July 1808, Hinckley to Melbourne Common turnpike Minute Book 6, 20 August 1823; W. Albert, *The Turnpike Road System in England 1663–1840* (1972), 85–7.

48. Griffin, *op. cit.*, 792–3.

49. Leicestershire County Record office, Ashby to Leicester Turnpike Road Minute Book, March 1795.

50. *Ibid.*, January 1806.

51. C. Wilson, 'Transport as a factor in the history of economic development', *Journal of European Economic History* (1973), 330.

52. Throsby, *op. cit.*, 103.

6 The Swansea Canal

H. POLLINS

ALTHOUGH it is true that the really large economic developments in the Swansea district came with the manufacture of steel and tinplate[1] and the mining of anthracite[2] in the second half of the nineteenth century, its industrial history is much longer. Indeed the contention of Sir Frederick Rees that the term "Industrial Revolution" is more appropriate to Wales than to the rest of the British Isles[3] requires some modification in the case of South-West Wales. There the mining of coal was old-established and copper smelting had begun in Elizabethan times. Nevertheless, before the nineteenth century developments were slow, and in 1801 Swansea had only some 6,000 inhabitants and was looked upon as a tourist resort as much as an industrial centre.[4] The growing export of coal on the one hand and the import of ores for the expanding copper industry on the other at the end of the eighteenth century both required improved port facilities and the development of communications from inland to the port, if Swansea was to become an industrial town. The Swansea Canal did much to meet these needs, and greatly facilitated the growth of Swansea as a major port of South Wales during the first half of the nineteenth century.

The need for improved transport from the pits to the port was indeed obvious. A large proportion of Swansea's trade was with the south-western counties of England—coal from Swansea in exchange for copper ore.[5] (As we shall see, petitions in favour of the Canal were received by the Commons from ports in Devon and Cornwall.) One of the reasons given for the promotion of the Canal was the fact that "vessels often wait several weeks for a turn to load, and are then obliged to proceed to other ports in search thereof".[6] There was no smooth and continuous supply of coal waiting to be loaded on the ships bringing copper ore or ballast, and of course a canal would assist the exploitation of iron ore, ironstone, coal, and limestone as well as extend the area of cultivation by carrying "such Limestone, and other Manure".[7] Some evidence that the most northern parts of the valley were developing is contained in the comments on the 1801 acreage returns. Thus of Ystradgynlais and Cilybebyll it was reported that: "Agriculture is very much upon the decline. The country in general abounding in minerals, commercial adventures have monopolized the lands formerly under corn for the use of

[1] C. R. Fay calls the period 1855–1900 "the age of steel, when Middlesborough, Sheffield and Swansea boomed": *English Economic History mainly since 1700* (1940), 113.

[2] D. Trevor Williams, *The Economic Development of Swansea* (1940), 57.

[3] J. F. Rees, "How South Wales became industrialized": *Studies in Welsh History* (1947), 132.

[4] D. T. Williams, 174.

[5] W. G. Hoskins, *Industry, Trade and People in Exeter, 1688–1800* (1935), 105–6, 164; D. T. Williams, 34–9.

[6] Walter Davies, *General View of the Agriculture . . . of South Wales* (1815), II, 405–6.

[7] *Journals of the House of Commons* (hereafter referred to as *C. J.*), XLIX (1794), 152–3. (Petition for the Swansea Canal Bill.)

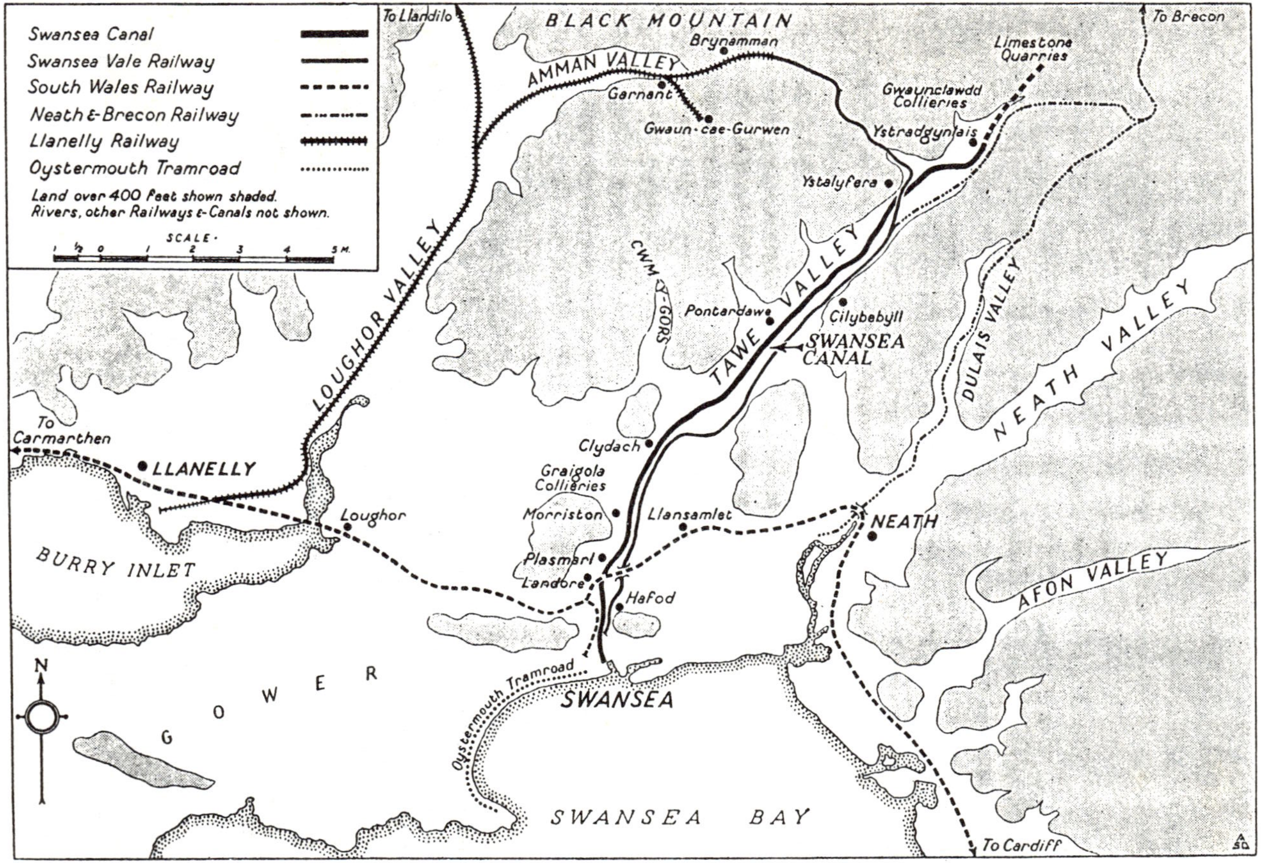
Swansea Canal
Swansea Vale Railway
South Wales Railway
Neath & Brecon Railway
Llanelly Railway
Oystermouth Tramroad
Land over 400 feet shown shaded.
Rivers, other Railways & Canals not shown.
SCALE
To Llandilo
BLACK MOUNTAIN
To Brecon
Brynamman
Limestone Quarries
AMMAN VALLEY
Garnant
Gwaunclawdd Collieries
Gwaun-cae-Gurwen
Ystradgynlais
Ystalyfera
LOUGHOR VALLEY
CWM-Y-GORS
TAWE VALLEY
Pontardawe
Cilybebyll
SWANSEA CANAL
DULAIS VALLEY
NEATH VALLEY
To Carmarthen
LLANELLY
Clydach
Graigola Collieries
Loughor
Morriston
Llansamlet
NEATH
BURRY INLET
Plasmarl
Landore
Hafod
AFON VALLEY
Oystermouth Tramroad
GOWER
SWANSEA
SWANSEA BAY
To Cardiff

their horses, and by the temptation of large wages induce the husbandmen to abandon cultivation for the hazardous employment of colliers, miners, furnace, forge and copper men." And also in Llansamlet (near to the town of Swansea): "a large coalery is opened in the parish".[1]

In general the place of the South Wales canals in the economic history of that region is well known and needs no emphasis. "Like many small unimportant canals and navigations round the coasts, their object was merely to bring an upland district into touch with tidewater; but their upland district was not some agricultural or secondary manufacturing region. It was trenched by the coal and iron valleys of Monmouth and Glamorgan."[2] The period of the Swansea Canal (1790s–1860s) was the period of the expansion of the copper industry, mainly around Swansea and along the Tawe valley.[3] From the 1840s and 1850s tinplate works and ironworks were built, and with the coming of steel in the 1870s and 1880s, together with the expansion of coal production, the industrial pattern of the valley was set, and remained substantially the same until the Second World War.[4]

On the whole therefore we may take for granted the place of the Swansea Canal in the economic history of the area before the coming of the railways in the 1850s, for clearly the pits, ironworks, and copperworks at the head of the valley and along the river would have been greatly retarded without it. This paper will be primarily concerned with the history of the Company rather than with the Canal's economic effects (if only because details of the traffic on the Canal have not survived).[5]

I

The actual dating of the Canal is of some interest. Its Act was obtained in 1794; like most canal and railway companies this company was effectively promoted during a boom, and we must look to the financial and commercial conditions of the economy in the early nineties for the reasons which brought it into existence. But there were other reasons too. The growing coal trade required improvements to the Swansea harbour, and throughout the eighteenth century various improvements were made. These improvements, however, whether carried out by the Corporation or by individuals, were on a small scale, and in 1791 an Act was obtained for a large-scale re-

[1] David Williams, "The Acreage Returns of 1801 for Wales", *Bull. Board of Celtic Studies*, XIV (1950–1), 60, 140, 141.

[2] J. H. Clapham, *An Economic History of Modern Britain* (1930–8), I, 78.

[3] J. F. Rees, 139. After the 1860s the copper industry declined.

[4] For more details see a forthcoming work, to be published by Watts & Co.: T. Brennan, E. W. Cooney, and H. Pollins, *Social Change in South West Wales* (provisional title).

[5] I should like to thank the British Transport Commission for permission to examine and quote material in their possession. I am indebted to the following for providing information and criticism: Mr W. C. Rogers (Estate Office, Swansea County Borough); Mr R. O. Roberts and Mr W. E. Minchinton (University College of Swansea); Mr L. S. Pressnell (University College, Exeter); Mr S. A. Broadbridge (London School of Economics); and Mr E. Brooks (St John's College, Cambridge). A summary of some of the points in this article appeared in *Gower* V (1952), 19–21.

modelling of the harbour, together with the erection of lighthouses,[1] and this was an incentive for the building of a canal from inland to the harbour. But even before 1791 a canal was being considered. In 1780 and 1790 surveys were made, but the former year was a war year, with high rates of interest; and it is known that in March 1793 the Corporation officially heard of the canal scheme, and decided to hold public meetings to arouse support.[2] This does not mean, however, that nothing was being done between 1790 and 1793. Perhaps surveys were being made, and people were becoming interested in the idea. Possibly too the successful promotions of other canal companies in 1792 and early 1793, before the boom burst, may have encouraged the promoters to go ahead on a scheme which, because of its local nature—it was not part of a trunk scheme—might have had difficulty in raising the necessary funds for an application to Parliament. However, we know little of these things.[3]

There is little doubt that once the decision had been made early in 1793 to go ahead with the scheme, the Canal Company proceeded in the "traditional" manner of company promotions in that period, i.e. public meetings were held, a survey was made, and subscriptions were taken.[4] In the original petition for the Canal it was stated "that several Meetings of the Proprietors of Lands . . . of the Members of the Corporation of *Swansea*, and of Persons interested in the Trade and Commerce of *Swansea*, have been held at the Town Hall at *Swansea*, pursuant to Public Advertisements". And after several such meetings it had been "unanimously agreed to petition Parliament . . . and an adequate Subscription was entered into".[5]

It is likely that the chief promoter was Rhys Davies, a local solicitor and owner of various valuable mineral properties in the Swansea valley. (His name appears on the manuscript material in the House of Lords Record Office, and he appeared before the House of Lords Committee.)[6] It was stated at the time that of 88 subscribers only nine had property on the line and "the rest may therefore be supposed speculators in the hope of a large unlimited interest".[7] But this was probably the exaggeration of opposition. The Subscription Contract of 1794, which is in the House of Lords Record

[1] W. H. Jones, *History of the Port of Swansea* (1922), 54. Such a scheme had been talked of from the late 1760s.

[2] The Corporation voted ten guineas for a survey, and also took ten shares of £100 each: D. T. Williams, 48. It is interesting to observe that other unreformed Corporations supported canals financially: C. Hadfield, *British Canals* (1950), 33, and S. G. E. Lythe, "The Canal Mania in East Scotland", *Scots Magazine* (May 1940), 119.

[3] There was apparently a contemporary scheme to build a canal from Swansea, through Neath, up the Neath valley. This may have kept alive the idea of a Swansea Canal. J. Phillips, *General History of Inland Navigation* (1792), 275.

[4] Hadfield, 34, for a generalized statement of canal promotion. Also G. H. Evans, *British Corporation Finance, 1775–1850* (1936).

[5] *C. J.*, XLIX, 153 (11 Feb. 1794).

[6] House of Lords, Committee Book, 2 May 1794 (House of Lords Record Office).

[7] *Case of the Duke of Beaufort and other Noblemen and Gentlemen on the proposed Swansea Canal* (?1794), in Swansea Public Library.

Office, contains a list of subscribers, showing 89 holdings of which 4 are probably duplicate.[1] Of these 85, it has been possible to identify about 70 as being local worthies who are well known in the history of the district.[2] Some of the remainder have com mon Welsh names and it has not, therefore, been possible to establish their identity. In the Book of Reference of the Swansea Canal (at the House of Lords Record Office), only 20 landowners are shown as holding land on the line of the Canal; even if they had all subscribed, there would still have been some 70 shareholders who were not landowners, but it would not follow that they were therefore speculators. The Subscription Contract may be compared with other lists of shareholders. In the Minutes of the General Assembly of Proprietors the names of those who attended proprietors' meetings are given, and in the early years a large proportion attended. Two meetings have been selected to see how far the shareholders of 1794 were stable. At the General Assembly of 2 September 1794 (the first after the Act was obtained at which such a list is given), there were 54 attendances (including proxies). Of these, 52 were in the Subscription Contract. At the Assembly of 4 July 1797, out of 74 names (including proxies), 54 were in the Subscription Contract. In addition another six may have been the children or other near relations of original subscribers.[3] In another shareholders' list, dated June 1872, out of 117 shareholders holding 533 shares, 63 holding 347 came from the Swansea district (together with a few from the other Welsh counties). All this suggests that the main financial sources were local, and that there were few, if any, speculators. It may well be argued that while this canal was promoted during a mania period (albeit towards the end of the mania), the smallness of the scheme and its unimportance to the rest of the country, and the fact that speculators probably were active mainly in the large towns, may have kept it clear of them.[4] Nevertheless the influence of a boom period can be seen in the statement that, at one public meeting, £40,000 was subscribed—it was obviously easier to obtain subscriptions (or rather promises to subscribe) in a boom.[5]

The shareholders and promoters appear to be divided into two main groups: those in the town of Swansea who wanted the town and port developed; and the owners of lands, collieries, and works in the valley who wanted improved communications to the port.

Progress was not straightforward. Two of the most influential local groups were in opposition—the Duke of Beaufort, the largest landowner, and the copper-smelting firm of Morris and Lockwood, tenants of the Duke. In the original plan for the Canal

[1] Such names as David Davies and Thomas Williams may, or may not, be duplicates.

[2] No addresses or occupations are given in the Subscription Contract.

[3] 1797 is chosen because it was probably the year when the last call was made (see below). Those who remained shareholders up to then were obviously not speculators.

[4] I am greatly indebted to Mr W. C. Rogers for identifying the names in the MS. Subscription Contract. The 1872 list is printed with the last *Report* of the Swansea Canal Company. It is in the custody of the British Transport Commission, and appears to be the only surviving *Report*.

[5] *Case of the Duke of Beaufort.*

it seems to have been proposed that use should be made of a private canal controlled by Morris and Lockwood. It was said that after a town meeting in favour of the Swansea Canal, Mr Morris insisted on the following clause being inserted in the proposed Bill:

> "That no Coal but Stone Coal, for the purpose of drying malt, or Culm, the small of such Stone Coal, for burning lime (except such caking or binding Coal as shall be used in any Iron Furnaces now built, or to be built, in any part of the Tawe Vale) shall pass along such part of the intended Canal as is now held by Thomas Lockwood and John Morris, Esqrs. by virtue of a lease under the Duke of Beaufort, without first obtaining the consent of the said Duke and of his lessees, Thomas Lockwood and John Morris. . . The like clause on behalf of Lord Vernon and Mr. Smith, in case the Canal should, in future by any powers of the Act, be brought over the lands in their possession between Landor and the Havod, in Swansea."[1]

The Canal Company claimed that their line had been agreed to by most of the landowners and occupiers (as required by Parliamentary Standing Orders):

	Assent	*Dissent*	*Neuter*
Landowners	24	3	2
Occupiers	69	6	6

Those who assented owned 15½ miles of land on the route of the Canal, the opposers five-eighths of a mile, and the remainder three-quarters of a mile.[2] (In the Book of Reference, 20 landowners are shown, of whom 18 assented, and 2 were "neuter"; 67 occupiers, of whom 59 assented, 3 dissented and 5 were "neuter". These figures are different from those just quoted, no doubt because the House of Lords documents refer to the final plan and survey, when most of this discussion had been concluded.)

The Duke of Beaufort and other opponents of the Canal put their case as follows:

> "This Canal is proposed to be carried near a mile and a half over an open colliery, held under a grant from the Duke of Beaufort's ancestors, and now in the possession of Lord Eliot, Mr. Lockwood, Mr. Morris and Mr. Calveley Cotton, and worked by several expensive steam engines, the ground of which (where the intended Line runs) has heretofore cracked from the mine to the surface, and if the same accident should again happen, this colliery, in which concern above £100,000 is embarked, would be lost, the great copper manufactories adjoining, in which also there is a capital of above half a million, would be rendered useless, and Lord Eliot, Mr. Lockwood, Mr. Morris, and Mr. Cotton, be liable to very heavy penalties for not supplying the coal contracted for."

[1] *Swansea Canal Case* (?1794), Swansea Public Library. This is also given in G. G. Francis, "Materials and Illustrations for a History of Swansea, Broadsides relating to Swansea and Glamorgan", fol. 46 (Royal Institution of South Wales, Swansea).

[2] ibid.

In addition the Duke of Beaufort was then receiving £2,000 per annum for "liberties to work coal" and wayleaves. The Canal would destroy the wayleaves. The new line of canal would take water from the canal belonging to "Lord Eliot, etc." and "their communication between their pits and engines will be cut off, and their various *rail* and *coal roads* intercepted in a manner fatal to the continuance of their concerns".[1] The Canal would also extract water required by the Swansea Harbour Trust who intended improving the harbour and making a floating harbour, so that there should be an "uninterrupted navigation for ships to the collieries and great copper manufactories".

The Duke of Beaufort and others, while not opposed to the idea of a canal, preferred it to be administered by commissioners, with limited dividends, rather than by a company. And if a good line were adopted and private property were secured from injury, "Lord Eliot, Mr. Lockwood, Mr. Morris and Mr. Calveley Cotton, will head a Subscription with £5,000 as well as make a present of their Canal to the public".[2]

The Canal promoters, for their part, were sure that the opposition to their Bill was based on two things. In the first place it was an attempt to keep the price of coal coming to Swansea high, and secondly, the suggestion that the Canal should stop short of Swansea was really a desire to build a new town at Plasmarl.[3]

This controversy was continued while the Bill was before Parliament. The petition was presented on 11 February 1794 and the Royal Assent was given on 23 May 1794. The discussions at Swansea were reflected in the rival petitions sent to the House of Commons:

In favour

24 Feb. Mr Rhys Davies; Mr Thomas Lane.[4]
10 Mar. The Mayor, Aldermen, Burgesses and principal inhabitants of Barnstaple, Devon.[5]
12 Mar. Gentlemen, freeholders and other inhabitants of Brecon.
The Portreeve, Aldermen and Burgesses of Swansea.[6]
13 Mar. Gentlemen, freeholders and other inhabitants of Great Torrington, Devon.[7]
18 Mar. The Mayor, Justices, Aldermen etc. of St Ives, Cornwall.[8]
27 Mar. John Bennet Popkin of Kittle Hill, near Swansea. (A land and mineral owner.)[9]
28 Mar. Owners of tin and copper mines in Cornwall.[10]
3 Apr. Mayor, Burgesses, Shipowners etc. of Marazion.[11]

[1] *Case of the Duke of Beaufort.* [2] ibid.
[3] Francis, fol. 48 ("Swansea Canal Bill"). In the 1790s John Morris was having built the planned village of Morriston; the plans for the canal presumably affected this.
[4] *C. J.*, XLIX (1794), 219–20. [5] ibid., 310–11. [6] ibid., 319. [7] ibid., 327.
[8] ibid., 347. [9] ibid., 381. [10] ibid., 388. [11] ibid., 414.

Against

17 Mar. Duke of Beaufort; Thomas Lockwood the Elder, John Morris, Revd. Edward Lockwood, Henry Calverley (*sic*) Cotton, and Thomas Lockwood the Younger.[1]

28 Mar. Duke of Beaufort; Swansea Harbour Trustees; Landholders, Manufacturers and others of Morriston. (These are largely those signing the petition of 17 Mar.)[2]

The expense of a Parliamentary contest, however, was spared, for the Duke of Beaufort and the other petitioners against the Bill declined to be heard by the Commons Committee.[3] Instead, some deviations proposed by the Duke[4] were accepted by the Committee, after the subscribers, landowners, and surveyor had agreed to them.[5] Some verbal amendments were made by the Committee, and some also by the Lords Committee, and the Bill finally passed.[6]

The Act of Incorporation (34 Geo. III, local Act, cap. cix) allowed the Canal Company to raise £60,000 in £100 shares and to raise a further £30,000 by shares or mortgage if this were considered necessary. The only unusual power authorized by the Act was that part of the Canal was to be built by the Duke of Beaufort at his own expense. This was called the Trewyddfa Canal; in return the Duke was to participate in the total revenue.

II

The Canal was built between 1794 and 1798. It commenced at Brewery Bank near the Strand in Swansea and terminated at Ystradgynlais in Breconshire; a tramway completed the route to the Hen Neuadd limestone quarries.[7] It had 36 locks and rose nearly 400 feet in its 16 miles. Little is known about the actual construction, as the Minutes of the Canal Committee are missing for this period. They begin in 1802.[8] But it is possible to obtain something from the Minutes of the General Assembly of Shareholders, which was usually held annually.

[1] *C. J.*, XLIX, 343–4. This reprints, in substance, the *Case of the Duke of Beaufort*. [2] ibid., 390.

[3] ibid., 401 (31 Mar. 1794). The Bill was read a second time on 28 Mar. after an amendment that the Bill be read "upon this Day Six Months" (i.e. rejected) had been withdrawn: ibid., 390.

[4] ibid., 425 (7 Apr.). [5] ibid., 444 (9 Apr.).

[6] ibid., 481, 496, 563, 622 (17, 28 Apr., 8, 23 May).

[7] J. Priestley, *Historical Account of Navigable Rivers, etc.* (1831), 611–12. *The New Swansea Guide* (1823), 82, describes the Canal as terminating near the Lamb and Flag public house (Ystradgynlais), notorious as the house where "the celebrated Hatfield, who seduced the maid of Buttermere, was apprehended".

[8] These documents are in the custody of the British Transport Commission. The report of the Committee meeting of 4 July 1794 is inserted in the volume of Committee Minutes commencing 1802. The Committee consisted of all shareholders holding 5 shares or more (as authorized by section 57 of the Act). At the first meeting of the General Assembly 56 names are given as members of the Committee. This method of appointing a Canal Committee was also employed by the Kennet & Avon Canal: *Felix Farley's Bristol Journal*, 26 Apr. 1794.

A survey had been made by Edward Martin in 1790,[1] but the engineer in charge of construction was probably Thomas Sheasby, a local mining engineer. The plans deposited by the Canal Company in 1793 and 1794 were drawn up by him, and he assisted the discussion through Parliament.[2] He also made a survey to alter the course of one of the streams feeding the Canal.[3] In addition at the General Assembly of 7 July 1795 Charles Roberts was appointed consulting engineer for thirty-nine weeks at twelve guineas per week, plus expenses.[4] It is not clear when the work of construction actually began. The Act was obtained in May 1794, the line was re-surveyed in August 1794, and a year later there is said to have been a rumour that the company was going to build a railway instead of a canal.[5] Work probably began in 1795, for in the Minutes of the General Assembly of that year there is a reference to the first call having been made on the shareholders (of whom eight had defaulted). In the same year just over four acres of land were sold to the Canal Company by the Corporation of Swansea for £2,000.[6] However, it was, apparently, finished by October 1798[7] and even before that date parts of it which were completed were open to traffic—in 1797 tolls were reduced, and a year later it was decided that in future tolls be collected monthly and the money laid out in completing the canal.[8]

It was decided at the first meeting of the Committee not to employ a contractor but to build the Canal under its own direct supervision.[9] It seems that the Company employed the labourers directly rather than through master craftsmen, which would have been the normal practice; in 1795 it was decided that "for the future no more than One shilling and sixpence per day be allowed to Cutters and Labourers for their Subsistance at the discretion of the Committee".[10] Apart from the canal and the tramway the Company built wharves (seven at £20 each) and owned barges. In 1797 four of the

[1] Jones, 106. He promoted the Oystermouth Tramroad.

[2] The plans are in the House of Lords Records Office and the Glamorgan County Record Office. *C. J.*, XLIX, 219, 444 and MS. Lords Committee Book, 2 May 1794, for Sheasby's activities before Parliamentary Committees. Priestley, 649, says he was the Canal's engineer.

[3] General Assembly, 3 July 1798.

[4] He had made a re-survey of the Canal in Aug. 1794. He was appointed for this purpose by the Canal Committee (Committee, Minutes, 4 July 1794). A copy of this survey dated 15 Aug. 1794 is in G. G. Francis, op. cit. He suggested the Canal should stop at Pont Gwaunclawdd, half a mile below Hen Neuadd, with tramways to complete the route, at a total cost of £45,720. This is probably the first mention of tramways.

[5] Jones, 114.

[6] *Report of the Estate and Property of the Corporation of Swansea* (1851), 45–6 (Library of the University College of Swansea). The date of the sale is given as 5 Nov. 1795.

[7] J. Phillips, *A General History of Inland Navigation* (4th ed., 1803), 590.

[8] General Assembly, 4 July 1797 and 3 July 1798.

[9] Committee, 4 July 1794. The Oystermouth Tramroad, although smaller, was built by contractors: C. E. Lee, *The First Passenger Railway* (1942), 9.

[10] General Assembly, 7 July 1795. The near-by bridge at Neath (1793) was also built by direct labour: *Trans. Neath Antiq. Soc.*, 2nd series, V (1936), 88.

barges owned by the Company were ordered to be sold at not less than £76 each and the proceeds of their sale were to be spent in building new boats.[1]

The Canal was financed by the issue of £100 Ordinary shares. No capital accounts appear to have survived, but the original cost "as far as can be ascertained" was given by a Parliamentary Return of 1890 as £53,300 for that part of the Canal made by the Company—i.e. not including the Duke of Beaufort's Trewyddfa canal.[2] The capital was £53,300 when the Canal was taken over by the Great Western Railway in 1872. As was normal, calls were made as money was required for construction. Apparently interest on calls was paid for a time.[3] By mid-1796 60 per cent of the shares had been called and eleven shareholders were in default,[4] but a year later it was found that another £6,400 was needed to complete the work,[5] and it was decided to raise the money from the existing shareholders by a *pro rata* contribution of 13 per cent. The Company was not sure of its powers and a year later, at the General Assembly of 3 July 1798, it was decided to take counsel's opinion as to whether the holders of the original shares could be compelled to pay this 13 per cent, and it was further resolved:

> "That those who have subscribed or shall subscribe to the open Shares, and also those original Subscribers who have paid or shall pay the twelve and a half per Cent. on their original shares be allowed Interest after the rate of five pounds per Cent. per Annum on such new Shares and twelve and a half per Cent. from the time of their several and respective payments to the next Annual General Assembly to be held in July 1799."

Although the details do not appear specifically in the minutes, it seems that in 1797 and 1798 money was advanced, mostly by old shareholders apparently, at interest, which it was intended should be funded into shares later. Possibly the Company had called up 87½ per cent of the shares and expected to have difficulty in calling up the remainder without the extra incentive of 12½ per cent interest.

In 1803 it was decided to effect this transfer of debt into shares, doubtless in order to lighten the burden of interest. In May the Committee decided that "application be made to those subscribers of the 7th Novr 1797 who have not converted their money paid in to shares to do so before the next meeting".[6] Thirty-nine people refused to do this, and their £3,641 6s. 1d. (principal £2,908 16s. 11d. plus interest to 6 September 1803)[7] was to be repaid by borrowing £3,000 from the Company's Treasurers (Lan-

[1] General Assembly, 4 July 1797.

[2] Canal Returns, *Parl. Papers*, 1890, LXIV (C.6083), 46. This is also the figure given in G. G. Francis, *Charters Granted to Swansea* (1867), 273.

[3] General Assembly, 4 July 1797. Resolution that interest on calls was to cease from 30 June previous.

[4] ibid., 5 July 1796. One of these eleven was owed money by the Company. It had been decided at the General Assembly of 7 July 1795 that defaulters were to be fined £1 for every call not paid.

[5] ibid., 4 July 1797.

[6] Committee, 3 May 1803.

[7] The details are given in the Committee Minutes, 6 Sept. 1803.

deg and Du Buisson, local bankers); but these would not supply it, and the money was obtained from Messrs Thomas Lynch and Edward Hawkins (of the Glamorganshire Banking Company) who promised the Company £3,000 for 5 years at 5 per cent on the condition that they were appointed Treasurers of the Company.[1] This sum was repaid by the Company in three annual instalments in 1807, 1808, and 1809 out of revenue.[2] In 1809 no dividend was paid so that this debt could be paid off.

As will be seen, once the Canal was built little if any addition to capital equipment was made. Apparently no more shares were issued. Maintenance of the Canal was presumably financed from revenue.

III

Once the Canal was built it settled down to a fairly steady and prosperous career. While it did not reach the heights of the Oxford or Kennet & Avon Canals, which usually paid over 20 per cent, and certainly came nowhere near the fabulous 75 per cent and 87⅓ per cent of the Trent & Mersey and Loughborough Canals,[3] it nevertheless paid 10 per cent or more in 47 out of 69 years as a separate dividend-paying body (1804–72). It is not known why dividends were not paid until 1804, six years after the Canal was opened. Possibly most of the net profit was used to pay interest on the money advanced in 1797 and 1798. Its most prosperous years were the 1850s, when many coal pits, tinplate works, and ironworks were constructed, and their production was growing, but before the railways came to extract the traffic.

Progress was generally upwards, but as the table of dividends shows there were setbacks. The normal fluctuations of the trade cycle accounted for some of this; the bankruptcy of Messrs Mercer, Ely and Treacher, probably the biggest traders on the Canal, in 1831, may have been the main reason for the decline in 1831 and 1832.[4] The fall in 1843 and 1844 was probably due to the building of the Llanelly Railway, which took some of the Canal's traffic, and the fall in the 1860s may all be attributed to the building of the Swansea Vale Railway almost parallel to the Canal. (On the other hand nothing is known about working expenses or the proportions, if any, carried over from one accounting period to another.)

When conditions were adverse, costs were reduced and so were tolls. In 1816 the wages of lock-keepers were reduced to 15s. per week (the lock-keepers to provide part of their equipment);[5] tolls were reduced in 1826, 1831, 1835, and 1843.[6] In 1846 an

[1] General Assembly, 5 July 1803.
[2] ibid., 11 July 1807; 5 July 1808; 4 July 1809.
[3] Hadfield, 158–9.
[4] Their bankruptcy is mentioned in the Committee Minutes of 6 Dec. 1831. The notice of the meeting of their creditors (*London Gazette*, 28 Oct. 1831, p. 2231), describes them as "Coal-Owners and Merchants". A year earlier their payment of tolls amounted to £4,200 (Committee, 7 Dec. 1830). No period is given but it appears to have been the amount paid in one year. In 1830 the total income from tolls was £8,740: *Swansea Directory* for 1830.
[5] General Assembly, 5 Mar. 1816. At the General Assembly of 11 July 1807 their wage had been fixed at 18s. per week.
[6] General Assembly, 4 July 1826; 4 Jan. 1831; 7 July 1835; 3 Aug. 1843.

economy committee was formed to reduce expenditure[1] and it was decided to reduce tolls in the same year.[2] In 1847[3] the Company decided to adopt the powers of the Canal Tolls Act, 1845 (8 & 9 Vict., cap. 28), which allowed canal companies "to vary their tolls and also from time to time to reduce and again advance them without exceeding the amounts originally authorized".[4] (The main point of the Act was the permission given to vary the tolls charged on different parts of canals.) An attempt to raise tolls in March 1849 was reversed when the traders complained of the poor state of trade.[5] And it is conceivable that the decision to sell the Canal office to the South Wales Railway in 1850 was influenced by the depression.[6]

When trade improved in 1853 the Company increased the tolls. The traders protested, and the Company accepted their complaints by reducing the tolls again; the Canal did not return the cash but recommended a dividend of 10 per cent, with the remainder of the net profit going for repairs and maintenance of the Canal.[7] There are unfortunately no data on which to base an assessment of the effects of these various decisions.

Under section 99 of the Swansea Canal Act railways or canals could be built to the Canal from the neighbouring pits and works. Those that were built seem not to have been built by the Company but by the various industrialists. Many applications were made to the Canal Company for such communications to be built, but invariably the Canal decided not to build and usually permission was given to the applicant to have them built himself.[8] The Canal Company, however, did not rest content with waiting for such applicants. In 1805 the engineer was ordered to make a survey of the country from the "head of this Canal" to the village of Trecastle, to "ascertain the practicability of endeavouring to form a junction with that part of the Country by Tram Road or any other way".[9] In 1840 the engineer, Mr Struve, was asked to make a branch from the Canal to the Burrows in Swansea, and also from the Canal to "terminate opposite the town".[10] In 1856 the Company considered extending the Canal into Mr James Palmer Budd's Ystalyfera iron works (built in the 1830s) and suggested that if he built the extension and junction (estimated cost £2,500 each), the Canal Company would allow him interest on the outlay at the rate of 7 per cent per annum so long as Budd's tolls on the Canal amounted to £3,000 per annum, or a proportionate sum if tolls were reduced. Budd, however, did not approve of the details and presumably nothing was done.[11] On the other hand four years later the Canal Company did build a communication between the works of Messrs Hallam & Co. and the Canal.

[1] ibid., 7 July 1846. [2] ibid., 1 Dec. 1846 [3] ibid., 5 Oct. 1847.
[4] Lord Hailsham (ed.), *Halsbury's Laws of England* (2nd ed. 1937), XXVII, 337.
[5] General Assembly, 6 Mar. 1849; 3 July 1849.
[6] ibid., 5 Mar. 1850. The Canal Company received £100 for it.
[7] ibid., 21 Mar. 1853; 20 May 1853.
[8] Like most tramways and wagonways in colliery districts these presumably were financed by the colliery (and works) owners. Cf. J. U. Nef, *Rise of the British Coal Industry* (1932), I, 386.
[9] Committee, 2 Apr. 1805. [10] ibid., 1 Dec. 1840.
[11] ibid., 4 Apr. 1846; 17 Apr. 1856; 9 May 1856.

The agreement was that Hallam was to pay the Canal 5 per cent for thirty years, the interest not to exceed £25 per year.[1]

The general history of the Canal was one of a steadily increasing trade with some attempt to tap more traffic sources by building or having built feeders from the works and pits to the Canal. Perhaps the largest feeder was a private tramway built in the 1820s. It has been described as follows:

> "About the year 1825, there was laid down by John Christie, Esq., of London, a long tramway, extending from the Gwainclawdd over the great Forest of Defynnock to Rhyd-y-briw, in the Vale of Usk. A branch, six miles in length, from Penwyllt to the head of the Swansea Canal, forming a junction with the main tramway, was also constructed.[2]"

The main activity of the Canal Company in its independent years was the administration and maintenance of the canal. Dividends were declared at the annual General Assembly held in July of each year. The dividends were normally payable towards the end of the calendar year, in November; when bonuses were paid in the 1850s it was usual to pay dividends and bonuses in two separate instalments (July and November). Traders were invariably allowed a credit of at least three months. Tolls could be paid in cash or by bills at a date of no longer than 35 days. (This might have been a useful form of short-term credit for the various works and pits.) The Company's turnover could not have been very large, so that the prompt payment of tolls was necessary not only to keep the Canal going but also to have sufficient to pay dividends when required. Thus in October 1826 (admittedly a period of poor trade), "a considerable deficiency appearing in the hands of Mr. Thomas Grove [the Treasurer] for the payment of the dividend of £12.10.0 per share on the 1st November next", it was resolved to call upon the traders to pay their arrears of tolls.[3] On the other hand there was a balance of £2,592 in the hands of the Treasurer at the beginning of September 1823 (a year of improving trade), which was expected to increase by the beginning of November.[4] Later, the Treasurer paid interest to the Company for balances of over £2,000; if the balance fell below £2,000 the Company allowed the Treasurer interest at the same rate.[5] One of the standard entries in the Committee Minutes is the need to remind traders to pay their tolls.

[1] ibid., 13 Apr. 1860.

[2] T. Jones and Sir J. R. Bailey, *A History of the County of Brecknockshire* (1909–30), IV, 92. In the Minutes of the General Assembly of 5 July 1825 there is a reference to John Christie, "renter, lessee or occupier" of mines in Ystradgynlais, who wanted to have a railway built to the Canal.

[3] Committee, 25 Oct. 1826. If necessary, the Treasurer was authorized to "give acceptances in the name of this Company at 2 months and to discount the same for any sum or sums not exceeding £1200" to enable him to pay the dividend.

[4] ibid., 2 Sept. 1823. The Committee thought the Treasurer ought to allow interest on a positive balance.

[5] General Assembly, 7 Sept. 1847. It is not known if this arrangement held before 1847. It may have originated in 1823 (see previous footnote).

The main fear of the Canal Company, of course, was the possibility of competition, but despite the railway mania of the 1840s and the obvious prosperity of the Canal in that period, it was not until 1860–1 that a railway was built up the valley. There had been tramways or wagonways connecting collieries with the river as early as the first half of the eighteenth century,[1] and there was the near-by Oystermouth Tramroad, famous as the earliest carrier of passengers,[2] so there was no complete lack of knowledge in the locality of the new means of transport. The first possible opposition was a railway proposed in 1830, whose promotion was doubtless influenced by the opening of the Liverpool & Manchester Railway in September of that year. Although it was advertised as intending to apply to Parliament for an Act, this "Swansea Railroad Company" did not proceed.[3] Unfavourable financial conditions may have retarded it as well as the technical difficulty—experienced also in 1840—of deciding on the actual site of the line; the plans of 1830 and 1840 suggested a line to the east of the river, whereas many preferred a line to the west.[4]

In the meantime the absence of communications acting as feeders to the Canal, or to the port of Swansea from the west, meant a loss of trade since the Llanelly Railway opened lines in 1840 and 1841 and attracted coal traffic from Cwmamman and Garnant, previously sent to Swansea.[5] People whose interests were centred on Swansea worried about this extraction of trade from that port, and proposed railways to connect the coalpits on the western side of the river with the port. Thus in 1841 John Henry Vivian, the Swansea copper magnate, was one of a group to form a company to build a line from Loughor to Swansea as well as a new dock at Swansea.[6] Apparently, this came to nothing, and Cameron's Coalbrook Railway, promoted for the same purpose, also was not built.[7]

These projected railways, since they would not take any traffic from the Canal, were welcomed by the Canal Company, even though they would not have added to the Canal's revenue. But proposals to build railways up the valley, parallel to the Canal and designed to take its place, were obviously different. Nevertheless the two schemes involving railways in the Tawe valley during the mania of the 1840s—the Welsh Midland and the Swansea Vale Railways—were welcomed by the Canal Company. The Canal Committee met a deputation (led by Joseph Bailey, M.P.) from the Welsh Midland Railway in 1845 and they agreed to recommend the Canal Company to lease the Canal to the Railway Company for 99 years from the date the railway was built. The details of the proposed lease—whose object was "to prevent the competition that must

[1] Jones, 261. For the history of Swansea railways generally see Jones, chapter 19. In the county of Glamorgan there had been wagonways from the seventeenth century: Nef, *British Coal Industry*, I, 386n.

[2] Lee, *First Passenger Railway*.

[3] *The Cambrian*, 13, 20, and 27 Nov. 1830.

[4] The cost of a line on the west was said to have been five times that on the east: Jones, 263.

[5] ibid., 263.

[6] ibid., 172 ff. *The Cambrian*, 23 Jan. 1841.

[7] Jones, 264. An Act was obtained for this latter railway but the line was not constructed: *Bradshaw's Railway Shareholders' Manual* (1854), 52.

otherwise arise"—are given in full in the Committee Minutes.[1] The lease was for £4,264 (or £8 per share), and "if the Canal Company should deem it advisable the Welsh Midland Railway Company would introduce a Clause into their Act reserving a given number of shares for the acceptance of the Canal Proprietors at the rate of four shares for one canal share or £200 in money at the option of the latter". Later the Welsh Midland Railway in conjunction with the Swansea Vale Railway proposed another agreement. The Swansea Vale Railway was to take over the lease as proposed by the Welsh Midland Railway and if the latter were to get their Act the Swansea Vale Railway would transfer their lease to the latter. Until the Swansea Vale Railway had exercised the option of taking the lease or opened the railway to the head of the canal, the railway would not take trade from the Canal above the Graigola collieries.[2] When the Welsh Midland Railway Bill did not pass Standing Orders and the Parliamentary Agents had decided that Parliament would not approve the proposed agreement with the Swansea Vale Railway, it was decided to oppose the latter,[3] and this opposition was continued in the 1847 Session (as well as opposition to the Swansea & Amman Junction Railway Bill).[4]

But except for a part of the South Wales Railway opened in 1850, no public railways were in fact built along the Swansea valley until the late 1850s, for the powers authorized by the Swansea Vale Rly Act of 1847 were unexercised—no doubt because of the financial difficulties of the late 1840s. During the next railway boom in the mid-1850s a new Act was obtained and the opening of this line on 24 December 1859 was the turning point in the history of the Canal.[5] The last attempt by the Canal to try to meet possible competition was the consideration by the Company of the desirability of building a branch line to Brynamman. Towards the end of 1860 the engineer reported on its importance[6] and the Committee appointed a sub-committee to obtain from Messrs Francis and Strick (of the Amman Iron Company) a guarantee that they would oppose the intended extension of the Swansea Vale Railway from Pontardawe to Brynamman, and to give the Canal the whole of their traffic for "say" thirty years.[7] A General Assembly was called, as was necessary, to consider the application to be made to Parliament but insufficient shareholders or proxies attended.[8] It was, however, accepted at a later General Assembly that a contractor should build the branch line (as required by the Amman Iron Company) and that he should work it on lease for twenty-one years,[9] but decision on a tender submitted by Mr H. Hutham was postponed for six months.[10] Eventually they decided to build the line and also a branch to collieries at

[1] 6 May 1845.
[2] Committee, 3 Feb. 1846. At first it had been intended to oppose the Swansea Vale Railway: General Assembly, 19 Dec. 1845.
[3] Committee, 9 Apr. 1846; General Assembly, 7 July 1846.
[4] General Assembly, 5 Jan. 1847.
[5] H. G. Lewin, *The Railway Mania and its Aftermath* (1936), 320, 482–3.
[6] Committee, 5 Oct. 1860.
[7] ibid., 7 Dec. 1860.
[8] General Assembly, 15 Mar. 1861.
[9] ibid., 2 July 1861.
[10] General Assembly, 7 Feb. 1862.

Gwaun-cae-gurwen in the possession of John Edward and Charles Morgan.[1] It seems unlikely that these lines were in fact built by the Canal Company.

The opening of the Swansea Vale Railway to Ystradgynlais in 1861 immediately reduced the traffic on the Canal and this was reflected in the dividends paid by the Canal Company. They fell from 18 per cent, paid for the twelve months ended 30 June 1860, to 9 per cent in 1863. After 1862 it paid 10 per cent only once again, in 1872. Even though it managed to retain the important long-distance trade of the Ystalyfera Iron Company, by charging special cheap rates (8d. per ton for 10½ miles), the traffic decline continued.[2] It was a most dramatic example of the effect of railway competition on a canal; and the difficulties of an industrial village which only had canal communications—described in a prize essay of the Ystalyfera *Eisteddfod* of 1860[3]—were quickly overcome when the railway was built.

The Canal was by no means finished; it was still paying good dividends, but the Company realized that its days were numbered, and the remainder of its history as an independent company consists largely of attempts to sell or lease the Canal to other railway companies. Between 1863 and 1866 it was negotiating with the Neath & Brecon Railway, and after that with the Great Western Railway.

In 1863 the Canal decided to approach the Neath & Brecon Railway for a lease of the Canal at a rent of 10 per cent. If this were not forthcoming the Canal would threaten to apply to Parliament in the next Session for power to build a railway up the valley.[4] In March 1864 the Canal decided to petition against a Bill of the Neath & Brecon Railway,[5] but the Canal, soon afterwards, approached the Railway for running powers over its proposed line. The Railway refused.[6] In October 1864 the Neath & Brecon, nevertheless, decided to acquire the Canal,[7] and after some negotiations a provisional agreement was made to lease the Canal for 999 years at an annual rent of £9,000 (this to exclude the Trewyddfa Canal).[8] A Bill was introduced in the 1865 Session but was withdrawn by the Railway Company after it had been decided that the Bill would be unlikely to pass.[9] In 1866 a similar Bill was again withdrawn by the Railway after some Canal traders had objected to certain clauses in the Bill.[10]

After the revelations of the peculiarities of the finances of the London, Chatham &

[1] ibid., 2 Jan. 1863. They had powers under section 99 of their Act to build such branch lines.

[2] ibid., 7 Oct. 1864; 1 May 1868; 2 Oct. 1868. The Canal Company's practice of charging lower rates for this "long trade", and higher rates for the carriage of goods from places nearer Swansea, was upheld in *Strick* v. *Swansea Canal Company*, 16 C. B. (N.S.) 243 (1864). Cf. the Furness Railway's cheaper rates for long hauls in the late forties: S. Pollard and J. D. Marshall, "The Furness Railway and the Growth of Barrow", *Journal of Transport History*, 1 (1953–4), 112.

[3] *Gardd y Gweithiwr* (*The Working Man's Garden*) (1861). Some relevant extracts, in English, are in D. T. Williams, 128–31.

[4] General Assembly, 6 Nov. 1863. [5] Committee, 4 Mar. 1864.

[6] Neath & Brecon Railway, Board, 3 May 1864. [7] ibid., 26 Oct. 1864.

[8] ibid., 2 Nov., 4 Nov., 10 Nov. 1864. Swansea Canal, General Assembly, 2 Dec. 1864.

[9] Neath & Brecon Railway, Directors' Report, 4 Sept. 1865.

[10] Neath & Brecon Railway, Board, 7 Apr. 1866.

Dover Railway, and the collapse of Overend, Gurney & Co., railway promotion schemes were temporarily out of favour. And the Canal's decision to build a line from Swansea to Morriston and Clydach[1] (to tap the growing passenger and freight traffic, which was a consequence of the growing importance of those places after the 1850s) came to nothing, probably because of the slump of 1867. In the meantime the Canal Company petitioned against possible competitors—the Swansea & Clydach Railway Bill, the Swansea Vale & Swansea Union Bill, and the Swansea Vale and the Swansea Valley & Harbour Junction Bills.[2]

In 1871, on the pretext of providing for the increased traffic from Clydach to Swansea, a sub-committee was formed to treat with the Great Western Railway, or some other company, to lease the Canal and lay down a subsidiary railway in connection with it.[3] After some negotiation, an agreement was made with the G.W.R. in November 1871,[4] and an Act was obtained in 1872, leasing the Canal to the G.W.R.[5]

The evidence placed before the Committee of the Commons which considered the Bill in 1872 consisted of the following main propositions:

1. The Canal had disadvantages; it became frozen in winter and in general did not allow a steady flow of goods to and from the various works.[6]
2. The coalfield needed developing.[7]
3. The Canal would profit by a railway because the district would develop and provide more traffic for the Canal.[8]
4. Better communication was required between Clydach and Morriston and Swansea because of the increased freight traffic to the port from those places, and also for the passengers going to Swansea on market days.[9]

Once the Act was obtained the history of the Swansea Canal as a separate concern was over, and its affairs became part of the general history of the Great Western Railway. As the figures show it declined steadily, and in few years made sufficient profit to pay the dividend provided for by the Act. For the G.W.R. exchanged £202 of 5 per

[1] Swansea Canal, Committee, 7 Dec. 1866; General Assembly, 2 July 1867. At this latter meeting it was decided to build such a line, to be financed by issuing the remaining 600 shares of the Company, to mortgage the Canal for £30,000, and to call in the sum of £3,750 due from the G.W.R. for land sold to that Company.

[2] Committee, 2 Feb., 17 Feb. 1866, 4 Jan. 1867.

[3] Committee, 5 Apr. 1871; G.W.R., Board Minutes, 13 Apr. 1871.

[4] G.W.R., Board Minutes, 3 May 1871; 15 June 1871; 2 Nov. 1871. Swansea Canal, General Assembly, 1 Dec. 1871.

[5] 35 & 36 Vict., local Act, cap. 152.

[6] Minutes of Evidence of the Select Committee of the House of Commons on the G.W.R. and Swansea Canal Company's Bill (1872) (in the custody of the British Transport Commission). Evidence of Richard Hughes (Manager of Landore Tinplate Works), Q.174, and John Cady (Manager of Williams, Forster & Co., owners of copper and silver works), Q.47.

[7] ibid., Evan Daniel (mining engineer), Q.229 ff.

[8] ibid., Joshua Williams (G.W.R. Manager in South Wales), Q.448 ff.

[9] ibid., passim.

cent cumulative preference stock for every £100 Canal share—ensuring therefore a virtually guaranteed income of 10 per cent to the shareholders. The Canal Company received £107,666 (original outlay £53,300) and the Duke of Beaufort received £40,000 (original outlay unknown).[1] Thus some £7,500 was needed annually to pay the dividend; in fact the Canal shareholders were subsidized by the railway more or less continuously from 1873. Even apart from this, the whole business must have been a loss to the G.W.R. It did eventually build a line from Swansea to Morriston, opened in 1881, but the Swansea Vale Railway, which tapped most of the traffic in the valley, was absorbed by the Midland Railway in 1876; and it was the Midland which finally extracted almost all the traffic from the Canal in the early 1900s by building branches from its Swansea Vale line to the various works and pits along the valley.[2] Thus the Canal is said to have lost some £2,000 per annum when the Graigola Colliery Company began to send its coal in 1905 by the Midland Railway, which built a branch direct to the colliery.[3] Indeed the motives of the G.W.R. in taking over the Canal are not clear. By 1872 the Canal's profits had been falling, obviously owing to the building of the Swansea Vale line, and it must have been generally accepted by 1870 that the canals taken over by the railways—mostly in the forties and fifties—were embarrassing to the railways.[4]

IV

The Swansea Canal was obviously extremely useful in the development of the Swansea valley, but it had the limitations of all canals. Above all, like most of those in South Wales, it had too many locks for the easy handling of traffic.[5] Its profitability depended not only on its necessity, but also, later, on the absence of competing railways. Once these were built the Canal quickly attached itself to a large railway company to avoid bankruptcy. By doing so its shareholders were provided with a steady income, but at the same time it proved a financial burden to the railway—like so many other canals and branch railway lines. Before the railways came, however, the Canal was the only means by which the Tawe valley could expand industrially, and the development of copper works and coal-mining along the valley was materially assisted by the Canal, whatever its technical limitations.

[1] Canal Returns, *Parl. Papers*, 1890, LXIV (C.6083), 46. It is unlikely that the 1½ miles of Trewyddfa canal cost £20,000 originally.

[2] *Second Report of the Royal Commission on Canals: Parl. Papers*, 1907, XXXIII, Part ii. Evidence of W. W. Grierson, chief engineer of the G.W.R., Q.23,532 ff. for a short history of the Canal after 1872. Details of the Midland Railway's interest in the Swansea Vale Railway are to be found in F. S. Williams, *The Midland Railway* (3rd ed., 1877), 309–16.

[3] *R.C. on Canals*: T. H. Rendell, goods manager of the G.W.R.: Q.23,727.

[4] Grierson, in his evidence (ibid., Q.23,532), said the Canal was in a prosperous state when the railway took it over in 1872. In a sense it was; the boom of the early seventies was enjoyed also by the Canal. But its subsequent history was not likely to be the same.

[5] E. A. Pratt, *A History of Inland Transport and Communication in England* (1912), 304.

SWANSEA CANAL DIVIDENDS, 1804–1872

Year Ending 30 June	*Dividend*	*Year Ending 30 June*	*Dividend*
1804	3%	1839	15%
5	4	1840	12½
6	5	1	15
7	5	2	15
8	5	3	14
9	nil	4	12
1810	8	5	15
1	10	6	16
2	10	7	10
3	10	8	10
4	10	9	9
5	10	1850	9
6	10	1	8
7	5	2	8
8	8	3	10
9	10	4	10 (plus bonus 10%)
1820	12	5	10 (plus bonus 6%)
1	10	6	10 (plus bonus 8%)
2	10	7	10 (plus bonus 8%)
3	10	8	18
4	11	9	18
5	14	1860	18
6	12½	1	13
7	16	2	10
8	15	3	9
9	15	4	9
1830	15	5	8
1	13	6	8
2	11	7	8
3	12	8	9
4	11	9	9
5	12	1870	9
6	13	1	9
7	15	2	10
8	16		

Source: MS. Minutes of the General Assembly of Proprietors of the Swansea Canal (British Transport Commission).

SWANSEA CANAL REVENUE ACCOUNT, 1872–1912

Period	*Receipts* £	*Expenditure* £	*Profit* (+) *or Loss* (−) £
7 months ending 31 Jan. 1873	6,670	2,450	+4,220
Year ending 31 Jan. 1874	14,468	3,443	+11,025
5	9,975	3,976	+5,999
6	8,457	3,379	+5,078
7	7,262	3,844	+3,418
8	7,597	4,698	+2,899
9	7,414	3,424	+3,990
1880	8,462	4,218	+4,244
1	8,482	4,159	+4,323
11 months ending 31 Dec. 1881	8,247	4,092	+4,155
Year ending 31 Dec. 1882	9,561	3,585	+5,976
3	8,958	3,573	+5,385
4	8,239	3,352	+4,887
5	8,027	3,804	+4,223
6	7,287	4,093	+3,194
7	6,626	3,872	+2,754
8	6,668	3,922	+2,746
9	7,661	3,890	+3,771
1890	7,903	4,423	+3,480
1	6,822	4,275	+2,547
2	7,091	4,266	+2,825
3	6,181	4,920	+1,261
4	5,157	4,198	+ 959
5	3,622	4,128	− 506
6	3,097	3,672	− 575
7	3,356	3,863	− 507
8	4,049	3,939	+ 110
9	5,034	3,520	+1,514
1900	4,814	3,578	+1,236
1	4,442	3,615	+ 827
2	4,727	4,134	+ 593
3	4,272	4,535	− 263
4	4,736	3,952	+ 784
5	2,577	2,834	− 57
6	2,625	2,723	− 98
7	2,705	2,881	− 176
8	2,580	3,191	− 611
9	2,640	4,140	−1,500
1910	2,534	4,110	−1,576
1	2,366	4,851	−2,485
2	2,089	4,470	−2,381

Source: Reports and Accounts of the Great Western Railway, 1873–1913. Up to Dec. 1912 accounts were made up twice a year. Under the Railway Companies (Accounts and Returns) Act 1911, the form of accounts was altered and the revenue accounts of the different canals were not shown separately.

7 *Josiah Wedgwood and canal management*

C. DAVIES

Few men did more than Josiah Wedgwood to encourage business and enterprise in the eighteenth century. His abilities as an entrepreneur in salesmanship and management led him to outstanding success in the pottery industry. He pioneered new techniques, working to maximise both profit and product quality by the strict organisation of his men and their methods. He insisted on punctuality, constant attendance, fixed hours, scrupulous standards of care and cleanliness, the avoidance of waste, and a ban on drinking.[1] But Wedgwood's interest in improvement went beyond his factory gates. As a young man he became involved in the promotion of turnpike trusts and, on a grander scale, of the Trent & Mersey Canal. In 1765 he devoted much of his time to the task of rallying support for that scheme in Staffordshire. He began the following year in London, lobbying MPs and giving evidence in favour of the canal to a parliamentary committee. In later life, as chairman of the concern in the 1780s, he made attempts to tighten lax organisation, and to put into practice some of the principles of regulation and supervision that had made his own business so successful.

The Navigation from the Trent to the Mersey was opened throughout in 1777. It was an impressive engineering achievement. It rose to 326 ft and crossed a major watershed. Its features included 109 road bridges and five tunnels, amongst them the extraordinary cut through Harecastle Hill. Its 163 aqueducts ranged in size from one of twenty-three arches across the river Dane to simple culverts for streams.[2] From the beginning it was intended to be a main line, a 'Grand Trunk', to which branches would be built and so connect together the manufacturing districts of the region. Nothing of its length had been built before in Britain, and only the Leeds & Liverpool Canal ever exceeded the Trent & Mersey's ninety-three miles.

Almost all canal companies adopted a similar management structure. The directors were elected by the shareholders to take policy decisions on their behalf. As they met in committee on only a handful of occasions throughout the year, they had to be well served by full-time executive officers. The business of the company was to maintain the canal as a public navigation. Tolls were charged and collected on goods which passed along the waterway. From this revenue the upkeep of the canal was paid, and the remainder went as profit in dividends to the shareholders. Day-to-day management was the firm responsibility of the officers. If the proprietors wished, their own role could be largely passive.

Tried and proved as this basic structure became, there are indications that its details were not entirely satisfactory in the early years of Trent & Mersey operation. No doubt there were initial teething troubles, and there were few places to look for example or guidance. The canal was exceptionally long, for the average length of waterways worked in England and Wales was never more than twenty-five miles, and half were shorter than sixteen miles.[3] It would not be surprising if problems were accentuated by the difficulties of ensuring that proper supervision took place over such a long distance. In addition it seems likely that the first canal directors assumed a greater entrepreneurial role than became usual. To encourage and profit from trade on the canal, the proprietors established their own carrying firm, known as Hugh Henshall & Co. The connections between the canal committee and the carrying business were intimate, and the managerial objectives and difficulties of the former presumably grew in accordance.

In the absence of company records (most were burnt in the 1940s) it is difficult to do more than speculate about such matters. But there is evidence that the company faced severe financial difficulties in its first years of trading. Construction began in 1766 and took eleven years to complete. The canal cost £300,000 to build, an enormous sum for the time, and investors hoping for a quick return on their capital must have been sorely disappointed. Some local trade commenced in the early 1770s, but the canal was opened throughout at a very difficult period. A general trade depression caused some navigations to lose traffic in the period 1775–79, and the situation worsened in 1780–84.[4] Against this trend the Trent & Mersey got off to a good start, revenue growing at an annual rate of about 20 per cent during the first four years of full operation. But it was not until 1781, fifteen years after authorisation of the works, that the first dividend of 2½ per cent was paid, and that was swiftly followed by a 13 per cent collapse in profits during the financial year 1782–83.[5] Shareholders may well have voiced their concern.

About this time two memoranda were placed before the canal directors: 'A few considerations respecting the offices of Auditor of our accounts, Cashier, & Banker, in order to discriminate their respective offices & duties'; and 'A few hints for the better regulation of our Canal boats & their Captains'.[6] It seems certain that they were written by Josiah Wedgwood, always a leading member of the canal committee and probably chairman at this time.[7] The memoranda are preserved in his personal 'Commonplace Book', along with documents relating to the problems of discipline and control which he had recently been tackling in his own works. Indeed, the style and descriptions used in the canal papers clearly reflect his business experience and terminology, rather than that more commonly associated with canal practice. It is not possible to say exactly why Wedgwood prepared these suggestions. The financial and other problems already mentioned may have been significant, and perhaps his own interests and enthusiasms also played a part. At a time when the company could not afford any slack the canal committee may have felt the need to examine their business minutely, to establish a clearly determined command structure, to regularise procedures and define the responsibilities of their officers, and to ensure that adequate supervision and disciplinary procedures were established. Wedgwood's first memorandum set out to distinguish the duties of the company's administrative and financial officers.

From the point of view of general administration, the auditor – known on other canals as the general manager, superintendent or agent – was to be the company's chief officer. His task was to safeguard their interests. Every bill had to pass through his hands, and a satisfactory explanation of its contents and justification supplied before he should approve its payment. If he was 'not in his own conscience satisfied with any explanation . . . he should lay the bill . . . before the committee, and take their instructions how he is to proceed'. In most cases he was expected to use his own judgement and bear a managerial burden, as well as to maintain the account books scrupulously. Part of his duties was to inspect the weekly returns of each wharf, 'giving occasionally praise or blame to the respective wharfingers as they shall seem to him to deserve; – and when any are faulty in this respect, and prove incorrigible to his admonitions, he is to report them to the committee'. The auditor had the special task of handling public relations, for the attitude he took could do most to encourage the first approaches of prospective customers, and ward off claimants. Letters to him were to be 'each replied to distinctly, in a clear manly stile [*sic*], equally distant from a pert forwardness, as from a fawning servility'. In his attention to detail Wedgwood seems to have revealed some of his own characteristics.

Less was said of the company's banker and cashier. The former – who was probably known as the treasurer elsewhere – merely had the task of safeguarding the takings, 'the office is so simple & well known'. Of the latter Wedgwood wrote, 'The whole of his business may be composed in a few words, – that he is to receive and pay all the company's cash – and to keep a simple cash account of all such receipts & payments.' The job required meticulous care and attention, but was otherwise straightforward and needed little initiative. The committee would even instruct the cashier when to pay his surplus into the bank.

'Let it be remembered,' Wedgwood thundered, 'that the single article of *Conveyance of Goods* is the ultimate object of our undertaking; And that the publick [*sic*] will be served, & the proprietors rewarded, in proportion as this branch is well or ill executed.' Clear and distinct guidelines for the company's officers were no doubt very necessary and useful, but Wedgwood's real interests lay elsewhere. He wanted the members of the canal committee to be able to play the full part of businessmen in charge of an important industrial enterprise. He had little interest in taking a back-seat role. Through Hugh Henshall & Co. the directors had a strong say in the carrying trade of the canal, and Wedgwood wanted to make their voice effective. Specifically referring to the 'problems of late', in his second memorandum he turned to devising means by which his part-time and often inexperienced colleagues could be kept fully informed, and up to date, with the detailed problems of the business. How were they supposed to know, for instance, whether there was evidence of misconduct at a wharf some forty miles from their offices in Stone? How could they anticipate developing problems before they came to a head?

Wedgwood's solution was for the committee to extend its own direct powers by employing 'walking surveyors' -- foremen, inspectors and troubleshooters for the canal directors.

> These officers have not yet received their instructions from the Committee, whose immediate servants they are, – and were, for good reasons upon their original appointment, made independent of all other control. They were indeed to be the Ears & Eyes for the Committee . . . and *their* minutes were intended for the inspection of the committee only. . . . This was intended to induce them to give their information and opinions freely & boldly, of whatever they saw amiss in things or persons; and they were made independent of all others, to prevent . . . their concealing anything from us for fear of other masters.

Their powers were to be wide-ranging. They had authority to investigate everything: to search around wharves and warehouses; to look at the various account books; to check and record what goods were lying where, and how long they had been so; to report on the condition of the company's various employees, overseers as well as labourers; to keep a look-out for all places where the canal might need repair; and, perhaps most important of all, to examine the company's boats as they passed, to check for signs of pilferage or damage, and to report on the state of sobriety of their respective captains. 'In short they were to inform the committee, by means of their minutes or otherwise, of everything which may be of consequence for the committee to know.'

The walking surveyors were intended to overcome the limitations of the canal committee, to ensure that they were never left out of touch or isolated, and to enforce an unprecedented degree of personal supervision and discipline. There is little doubt that they would quickly have become the most unpopular men on the canalside. As snoopers for the management they would have been hindered by silent obstruction at best, open to temptation, bribery or violence at worst. Wedgwood argued that the committee would have to go out of its way to compensate for these difficulties, and to encourage them to take a pride in their objectivity. They were to be independent, not set in any administrative hierarchy, and answerable only to the directors. Attempts to defraud the company would be discouraged by the constant inspections of the surveyors, while wharfingers and warehousemen would be encouraged to ensure the prompt despatch of goods. They were to have direct access to the directors and

> should be appointed by the committee alone. If they are properly attended to, and encouraged in their duty, they may be of singular service to the navigation; but if they are neglected by the committee, appointed & governed by the officers, they will want that power and authority which the committee alone can stamp their office with, will become insignificant & useless, and perhaps do more harm than good.

Nothing could do more to improve the efficient and rapid conveyance of goods than to regulate the passage of the boats themselves. Complaints about the pilferage of goods and the poor behaviour of boatmen were legion throughout the canal age; they gave the waterways a bad name in a way which poor administration could never rival. The proprietors of the Trent & Mersey Canal had a strong interest in ensuring that the employees of Hugh Henshall & Co. exhibited behaviour beyond reproach. Wedgwood suggested that the captain of each boat should keep up to date a 'way bill', in effect a log book similar to that used by a contemporary lorry driver, and not to be confused with the common canal trading document of the same name. The times of arrival and departure at various places were to be entered on this sheet, together with comments by wharfingers and walking surveyors who were to inspect the boat as it passed, check its contents and take disciplinary action when necessary. No matter where a boat was located, its captain could never avoid the possibility of inspection and due report to his employers. At the company's offices the auditor was to examine every waybill in detail: 'much depends upon this examination, and unless it is done with attention & effect, all that has been done stands for nothing'. The officer was to select items of interest and place them before the committee for further consideration. Wedgwood urged his colleagues to browse through recent waybills before their meetings, for detailed study would give committee members 'such a degree of accuracy & precision in the knowledge of our concerns, as may be very advantageous to themselves & the rest of the proprietors'.

To illustrate his suggestion, a specimen waybill was included in Wedgwood's memorandum (see appendix). A fascinating document, carefully prepared and of great interest, it was nonetheless almost certainly a fiction, designed to impress his fellow directors with the advantages of the scheme. The specimen detailed thirteen days in the working of 'Boat No. 1', commencing on the first day of January 1783. During this brief period the boat was involved in a rapid series of incidents which included damage to a lock, a drunken captain, the theft of cargo and eventual apprehension of the culprit, all of which were witnessed by thirteen company employees, an improbable half of whom had the same initials to their name.[8] Of incidental note is the extent to which accurate time keeping was needed to work the system. Had the idea been widely adopted by canal companies then, who knows, it might have induced pressures for the standardisation of church and public clocks throughout the country, and the introduction of a single 'canal time' some forty or fifty years before just one time appeared in the railway timetables.

Wedgwood had no doubt that his plan would fulfil every expectation, and would introduce a new element of discipline into the carrying business.

> From this examination it is, that proper objects will appear for praise or reward, for censure and for punishment; and without such discrimination, and consequent distinction made publickly, amongst the people employed by us, officers as well as labourers . . . there will be an end of all emulation to merit the good opinions of their employers . . .

The strict implementation of the waybill system would remove the grounds for many traders' complaints.

> When the captains are assured . . . that their conduct is strictly scrutinised, they will be doubly careful of their cargoes, – and of the boats, – the locks – or whatever they have to do with. . . . This regulation will, besides, introduce & keep alive such habits of attention & industrious execution, amongst our navigators of all classes, as will tend very much to ensure safety and despatch to the goods committed to their care, and consequently more tonnage to our canal. . . .

It would be easier to assess the significance of Wedgwood's various proposals if it were known whether or not they were ever implemented. His particularly innovative ideas seem to have been largely rejected. Figures with such romantic names as 'bank riders', 'knobstickers' and even 'walking surveyors' do occasionally appear in the literature of canals, but none was endowed with the authority Wedgwood would have wished. In March 1785, for instance, the General Assembly of (Trent & Mersey) Proprietors were presented with a 'Memorial' signed by fifty-one potters, complaining about inadequate facilities at Preston Brook: insufficient warehouses; too few porters to assist with transhipment; delays and damage to goods; all the sort of things which the committee's 'Ears & Eyes' might have revealed had they been effective.[9] As for the scheme to control boat captains, nothing whatsoever seems to have come of it.

It may well be that Wedgwood had a specific reason for the preparation of his memoranda. If so, it cannot have been sufficient to warrant attempting such novel changes. Shareholders' discontent was soon relieved. Afer 1783 the Trent & Mersey became increasingly successful, and the company's enormous dividends did much to encourage canal speculation. With his strong and dynamic personality, Wedgwood planned to give the directors broad executive and managerial powers. He seems to have wanted the committee to run the canal as he would his factory. It is doubtful whether his ideas received much support. Less interventionist members were probably satisfied with the work performed by the company's professional officers, and may have regarded their committee work as a necessary but tedious public duty. In any case, the 'walking surveyor' plan would likely have proved a failure, bitterly resented by every employee from chief officer to common labourer. It is perhaps best for Wedgwood's reputation that it was passed over.

His scheme for control of boats and their captains had merits, and might have silenced many of the canal system's later critics, but in its actual context was unduly rigorous and needlessly complex. Canal companies could have built up their own fleets and become both maintainers of the navigation and principal users of the waterway, on similar lines to railway companies in the nineteenth century. But that was not the case. From a sample of records for the year 1795, it has been suggested that as many as 80 per cent of boats may have been owned privately, not by canal companies.[10] As trade on canals developed, that pattern probably became increasingly marked. Canal companies did not go in for carrying in a big way. The regulation of boats and discipline of captains remained the responsibility of their particular employers, whether in the case of Hugh Henshall & Co., Pickfords, or of any other firm. To have been really effective, Wedgwood's plans would have required a degree of co-ordination and co-operation between canal owners and canal users which in practice was never achieved.

APPENDIX

Specimen way bill [log book]*
Boat No. 1. Capt. John Roberts

1783 Jany	*Places*	*Miles*	*Time am*	*pm*	*Remarks*
1	Shardlow	–	–	–	Arrived with 40 crates from Stoke. Good condition'd. – (signed) H. More.
2	"	–	–	–	Taking in loading.
3	"	–	7	–	Left – with 20 tons flint for Longport. (signed) H. More.
3		13	–	5	Met No. 1 with 20 tons. All well. (signed) J.B. surveyor.
3	Horninglow	15	–	6	Stopped all night.
4	"	–	7	–	Left – all well (signed) J. Percival.
4	Bramley	12	–	3	Passed – all well (signed) H. Lester.
4	Rudgely	4	–	–	Stopped all night.
5	Stone	15	–	–	Stopped all night – all well – Given 2 tarpaulins & a windlas. Repaired the helm.
6	"	–	9	–	Left – all well (signed) J. Myat.
6	Stoke	8	–	1	Passed – all well (signed) J. Burgess.
6	Longport	4	–	3	Arrived with 20 tons flint – all well. (signed) J. Badkin.
7	"	–	–	–	Detained here for want of loading.
8	"	–	–	–	"
9	"	–	10	–	Left – with 50 crates for Preston [Brook]. (signed) J. Badkin.
9	Wheelock	10	–	5	Arrived – all well – stopped all night. (signed) R. Caulton.
10	"	–	7	–	Left – (signed) R. Caulton.
10	M[iddle]wich	6	12	–	Passed – The Capt said he had stopt [*sic*] with a friend – he was a little in liquor – (signed) J. Biggs.

1783			*Time*		
Jany	*Places*	*Miles*	*am*	*pm*	*Remarks*
10		4	–	4	Met No. 1 without the captain – Lock damaged – and the boat a fresh bruise – (signed) J. Barnett, surveyor.
10	Wincham	2	–	–	Stopt [*sic*] the boat all night, the captain being much in liquor & the boat damaged – He is in general a sober man – (signed) J. Ball.
11	"	–	7	–	Left, quite sober, & promises to continue so. (signed) J. Ball.
11	Preston [Brook]	11	12	–	Arrived, with 50 crates – Two seemed broke open. Examined the captain, who said he did not know how it happened, but owned he was absent from his boat, & sent another man with it a few miles. He seems very sorry for what he has done, promises not to be guilty of the like again, and will endeavour to find out where the crates were broke open. I am persuaded from circumstances that the boy knows, but he will not confess at present: I should propose examining the boy at the committee – I have written to B[entley] & B[oardman] – at Liverpool, to examine the suspected crates, and I send a man back with the boat – The captain seems disposed to do all in his power to find out this theft, & I believe he will be more careful for the future. (signed) J. Wooley.
12	Preston [Brook]	–	10	–	Left – with 20 tons clay, for Etruria. (signed) J. Wooley.
12	Wincham	11	–	3	Passed this wharf – 20 tons clay – all well (signed) J. Ball.
12	M[iddle]wich	6	–	6	Arrived, & stopt [*sic*] all night – The man who robbed the crates is detected, & in custody. The goods are in the captain's possession – I have directed him to leave them at the general office.
13	"	–	7	–	Left – all well – (signed) J. Biggs.
13	Wheelock	6	10	–	Passed – all well. (signed) R. Caulton.
13	Etruria	14	–	6	Arrived, with 20 tons of clay – All well. (signed) E. Rochford.

*Some slight alterations in punctuation have been made for clarification.
Source. 'A few hints for the better regulation of Our Canal boats & their captains'. Josiah Wedgwood's Commonplace Book, vol. 1, p. 243, JWMS 39/28408.

Notes

1 N. McKendrick, 'Josiah Wedgwood and factory discipline', *Historical Journal* (1961), p. 38.

2 J. Ward, *History of the Borough of Stoke-upon-Trent* (1863), pp. 169–70.

3 Calculated from H. de Salis, *Handbook of Inland Navigation* (1904), using a sample of 162 canals and river navigations, with a total length of 3,907 miles.

4 J. R. Ward, *The Finance of Canal Building in Eighteenth-century England* (1974), p. 165.

5 Notebooks of Edward Sneyd of Lichfield, William Salt Library, Stafford, HM 37/40.

6 Josiah Wedgwood's Commonplace Book, vol. 1, pp. 241–7, JWMS E39/28408, from which all further quotations are taken unless otherwise stated.

7 It has not previously been established that Wedgwood was ever the chairman, but a document in the Wedgwood archives does appear to prove the case: 'History of a Defamatory Report', in Wedgwood's Commonplace Book, vol. 1, p. 115.

8 The initials 'J.B.' – the same as the canal's first engineer, James Brindley. I am indebted to Harry Hanson for pointing out that there was a real 'R. Caulton' at Wheelock (see appendix). In 1795 Richard Caulton owned three boats which worked from there, and he may well have been a wharfinger or agent for the canal company in the 1780s.

9 Committee of Commerce, Minutes 1784–1790, p. 25, JWMS E28404-39.

10 H. Hanson, *The Canal Boatmen* (Manchester, 1975), p. 9.

Acknowledgements

I am grateful to Messrs Josiah Wedgwood & Sons Ltd, Barlaston, Stoke on Trent, for permission to quote from the Wedgwood papers, and to Keele University Library, where the archives are deposited.

I would like to thank Theo Barker, Charles Hadfield, Harry Hanson and Jean Lindsay for their constructive comments.

8 *James Green as canal engineer*

C. HADFIELD

JAMES GREEN's work in canal engineering is of interest to us because he solved in terms of his medium the problem presented by the hilly country of the south-west of England. He did this by abandoning the large waterway with its locks, and substituting a narrow canal upon which moved tub-boats working in trains, each boat with a capacity of a few tons. Changes of level were overcome either by inclined planes or by lifts. Though his ideas were not new, he developed them more practically than his predecessors, and many miles of canal in Cornwall, Devon, and Somerset remain as his monument. This essay confines itself to describing Green's work in the south-west, and makes no attempt even to touch on his other achievements.

He was born in Birmingham in 1781, the son of an engineer,[1] with whom he worked till 1801, when he found employment with Rennie in "extensive surveys, canal works, the drainage of bogs and fens, and the design and execution of engineering works generally, both in England, and in Ireland". I do not know which canals are referred to here, but one of them was the Chelmer and Blackwater Navigation in Essex, the sea-lock of which he reconstructed.

In 1808 he was appointed Surveyor of Bridges to the County of Devon, following the tradition of Telford, who had been appointed twenty years before to a similar post in Shropshire. It was at that time the practice for those employed as county surveyors to undertake private work also, and this Green did after he had settled at Exeter. He drew attention to himself by the following advertisement:

> "James Green, Surveyor and Civil Engineer, begs leave to inform the noblemen and gentlemen of Devonshire and the adjacent counties that, in consequence of his recent appointment as general surveyor of the Devon County Bridges, he has taken up his residence in the city of Exeter, and respectfully solicits their patronage in the several branches of his profession."[2]

It seems likely that his first active canal job in the south-west was as engineer to the Exeter and Crediton Canal, which had been authorized in 1801, and which began cutting in 1810, only to be abandoned about a year later after half a mile had been excavated.[3] In 1817, however, his first important task came along, when he was employed to report upon the proposed Bude Canal.

[1] I have taken the biographical details from the *Minutes of Proceedings of the Institution of Civil Engineers*, ix (1849–50), 98 et seq.

[2] *Exeter Flying Post*, 10 Nov. 1808.

[3] An advertisement in the *Exeter Flying Post* of 13 June 1811 for 100 canal cutters is signed by James Green, and the only canal cutting at that time in the neighbourhood was the Exeter and Crediton. It is most unlikely that the first part of the Grand Western, with which as far as I know Green had nothing to do, is indicated.

Sand taken from the shore at Bude, because of its high lime content, had been used in the country round as a fertilizer since the Middle Ages. In March 1774 Edmund Leach had proposed a narrow and winding canal, ninety-one miles long, to join Bude with Calstock on the Tamar, mainly for the purpose of distributing this sand.[1] It was to have five inclined planes. An Act was passed to authorize a canal on the lines of this proposal in the same year, but no action followed. The idea was revived about ten years later, and in 1793 was taken up strongly by the third Earl Stanhope, who owned an estate at Holsworthy, and who was himself an inventor of parts. Leach's book had been published about 1791, and it included a detailed description of an inclined plane powered by water and moving boats on a cradle from one level to another. Lord Stanhope worked on the problems involved, and when Robert Fulton, who reached England at this time, heard of the project, he began a correspondence with him of a rather elementary kind upon inclined planes and lifts. Fulton had no practical canal engineering experience, and this correspondence[2] appears to have been the basis upon which he wrote his book, *A Treatise on Canal Navigation*, which was published in 1796, and advocated small canals and inclined planes.[3] The war caused the project to lapse once more, and it was not till after the peace that the scheme was again taken up, by the fourth Earl Stanhope, who in 1817 employed James Green and Thomas Shearn to make a survey[4] for a canal between Bude and Holsworthy. The report they made says:

> "In consequence of the resolution of your General Meeting, on the 13th August last, directing me to examine the country between Bude Haven and Holsworthy, with a view to ascertain the practicability of forming a Communication between the two places, by means of a Canal or Rail-Road, with a branch towards Tamerton Bridge;[5] taking, as a basis, the lines proposed by the late Earl Stanhope, and, as the leading object, the conveyance of sea sand into those districts which apply it as a manure... If... a canal be resolved on, it ought to be on a small and inexpensive scale; and some of the same reasons which operate against a Canal on Lock principles, form serious objections to the adoption of a Rail-Road, inasmuch as 362 ft of the whole elevation must be attained before it can reach the line of the Tamar, and that within a space of five miles and a half...
>
> "With this view of the subject, I have made my survey for a Canal... of small size (viz) 10 feet wide at bottom, 3 feet deep in water, and 19 feet wide at top. I

[1] E. Leach, *A Treatise of Universal Inland Navigation* (n.d.).

[2] See H. W. Dickinson, *Robert Fulton, Engineer and Artist* (1913), chs. iii and iv.

[3] I only know of one survey carried out by Fulton. This *Report on the proposed Canal between the Rivers Heyl* [*Hayle*] *and Helford* (1796) may possibly have been known to Green. It proposes inclined planes.

[4] *Report of the Committee appointed to conduct a Survey*... 15 Apr. 1818.

[5] On the River Tamar.

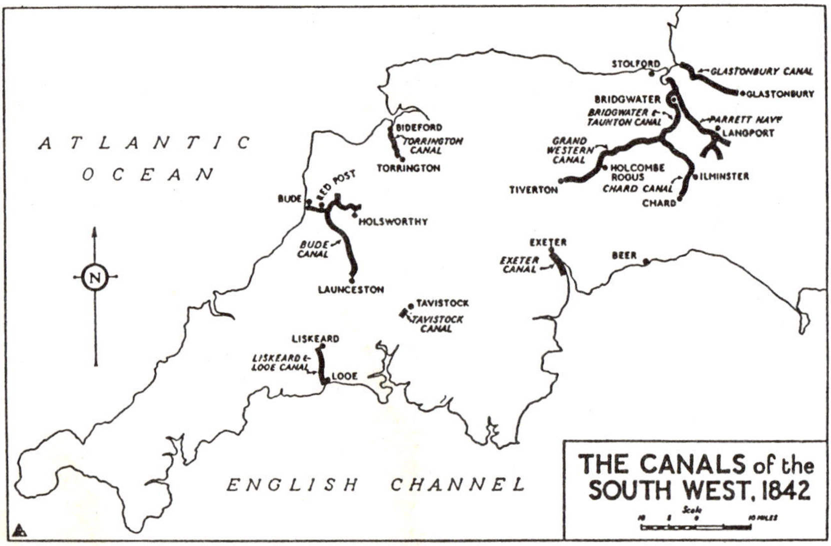

THE CANALS of the SOUTH WEST, 1842

propose that the boats shall be constructed to carry 5 tons each, and that one horse shall draw at least 4 of such boats.

"I propose to overcome the ascent from the sea to the summit-level, by means of inclined planes, up which these small boats may be readily passed by machines of a simple construction, moved by water. These machines may be executed at one-third the expense of locks, and worked with one-third the quantity of water which locks would require, with a saving of more than four-fifths of the time in passing them."

Here were all but one of the ingredients of Green's canal engineering thought. Small canals, tub-boats and inclined planes were not new. The first canal to be built in the West Country after the Exeter, the St Columb Canal,[1] was of this type, and may have had an inclined plane of a primitive kind by which goods were slid up and down a flat wooden slope from the end of the canal on the cliff of Trenance Head to the sea below. As we have seen, the same ideas were held by Leach, the third Earl Stanhope, and Fulton, while in 1817, the year of Green's survey, the Tavistock Canal had been opened. This was of tub-boat character, and had an inclined plane to connect the end of the canal with the river Tamar, built by a local mining engineer. On this plane, however, it was not possible to carry the boats themselves from one level to the other. Instead, the goods had to be moved by crane from the canal boats to the trucks that

[1] Authorized by Act of 1773, partially built soon afterwards.

ran on the plane, and again from them to the river barges.[1] Green must have been influenced in all these ways, but it is likely that his real instruction came rather from the tub-boat canals and inclined planes of Shropshire.[2] In his Report of 28 April 1829 to the Grand Western Canal proprietors, he suggests the use of inclined planes which could transport boats in cradles, and says that canals on this principle have existed for thirty years in Shropshire.

An Act was obtained in 1819, and by 1825 the Bude Canal was finished. It began with a sea-lock at Bude Haven which admitted ships of 300 tons to the first quarter-mile of canal which formed the basin. Sand was loaded into trucks on the beach, and these ran on rails over the foreshore and up an incline to the basin to be loaded into barges.[3] The railway was Earl Stanhope's idea, and superseded Green's original proposal to let the barges themselves rest on the sand at low tide and be loaded direct. The fifty-ton barges carried the sand for a little over two miles of broad canal to Marhamchurch, passing two locks on the way. At Marhamchurch the sand going further by water was transferred to four-ton tub-boats, which were fitted with small wheels of 14 in. diameter projecting from the sides. These wheels ran upon the rails of the inclined planes, and so saved the use of cradles or caissons[4] but they caused damage to the canal banks, and were not altogether satisfactory. They again probably originated in an idea of Earl Stanhope's.

There was a plane at Marhamchurch and another two miles further on at Hobbacott Down. This was the biggest on the canal, 225 ft rise and 907 ft long. At Red Post, beyond Hobbacott, the canal split into two branches, that to the east going to Blagdon Moor beyond Holsworthy and the reservoir at Alfardisworthy, with a plane at Veland, that to the west to Druxton near Launceston, with planes at Merrifield, Tamerton, and Werrington, a total length of thirty-five and a half miles. All these planes had water as their motive power, the means being a waterwheel except at Hobbacott Down, where two buckets were used in wells of depth equal to the height of the plane. In order to draw a boat up the plane, one bucket would be filled with fifteen tons of water and would then descend in its well, at the same time pulling the boat up the slope and the other bucket up its well. When the full bucket reached the bottom it hit a striker, which released the water into an adit leading to the lower pound of the canal. This machinery was a good deal more unreliable than the waterwheels, and a steam engine was held in reserve and was, indeed, often used.

The few original records[5] that remain do not reveal Green as taking an active part

[1] There are conflicting descriptions of this plane, but the account of Worth ("Early Western Railroads": *Transactions of the Plymouth Institution*, x, 1887–90) is by an eyewitness, and I have followed it.

[2] For a short description of the canals, see my *British Canals* (1950), 139–40; of the first plane to be built in Shropshire, ibid., 59–60.

[3] See *Railway Magazine* xl (1917), 96 and li (1922), 73.

[4] The boat rested dry in a cradle, or could be carried down floating in a tank or caisson.

[5] In the possession of the Bude Urban District Council and in private hands.

either in the construction of the canal and its planes, or in overcoming the many teething troubles, but Green himself in his Report of 28 April 1829 to the Grand Western proprietors definitely claims that he built the Bude Canal and its planes, and in 1831 he still had some position of responsibility in relation to the canal.

Between 1820 and 1827 Green was employed in enlarging and extending the Exeter Canal, and in building a basin at Exeter. This work was straightforward canal engineering, though his method of building Turf Lock is technically interesting, and I do not intend to describe it here.[1]

From 1769 there had been proposals to cut a canal across the peninsula of the south-west. The Grand Western Canal, partly built between 1810 and 1814 from Holcombe Rogus on the Somerset border to Tiverton, was part of one such scheme; also in 1811 Rennie had made a survey for a small ship canal across Somerset to cost £1,350,000 including a harbour at Beer. From this date onwards interest was maintained in the possibility of a canal that would enable shipping to avoid Land's End, and in 1822 Green was asked to report on the possibilities by a group of local people, among them Lord Rolle, whom we shall meet again. At this time efforts were being made to revive an old project of the canal mania, subsequently authorized in 1811, for a barge canal from Bristol to Taunton, and Green appears to have been instructed to survey for a barge canal to carry local traffic, rather than for a ship canal to carry through traffic. His proposal was for a small canal from the Bristol and Taunton near Taunton to Beer, 22 ft 6 in. at top and 3 ft 6 in. deep, except on the summit level of thirteen and a half miles, where it would be wider and deeper in order to provide a reservoir of water to work the planes. He proposed five of these, varying from 50-ft to 190-ft rise. He also suggested a pier at Beer, on to which barges should descend straight from the last incline by means of an extension to the rails of the plane. He therefore reported on 16 December 1822 that:

> "I am of opinion that a Canal on a very small scale, on which boats of four or five tons may be navigated in sets of four, six or eight boats, drawn by one Horse on the level Ponds of Canal, and passed singly up and down inclined planes by the aid of water Machinery, instead of locks, will best answer the intended purpose."

His estimate of cost, including the pier at Beer, was £123,156. A meeting was held at Chard on 18 December, and it was agreed to apply for an Act. This was not obtained, and the project is lost sight of after 13 February 1823, when Mr Nicholetts, an emissary of the promoters, appeared before the River Tone Conservators to ask whether, if the Bristol and Taunton Canal were not completed, Green's canal could lock into the Tone Navigation. The answer was "no".[2] Instead, the prosperous state of the country swung

[1] See *Memoir of the Canal of Exeter from 1563 to 1724, with a continuation from 1819 to 1830; and an appendix containing the Reports on the Canal by James Green: Mins. of Procs. of I.C.E.*, iv (1844–5), 90.

[2] Records of the Conservators of the River Tone, Taunton.

opinion back to the idea of a ship canal. Mr T. Pyke, who had been one of the earliest supporters of such a canal, published a pamphlet[1] in 1823 in which he advocated a ship canal, and referred to Green's project by saying that "a more ridiculous, ill-advised, inefficient, incompetent scheme was never devised. . ." The idea grew till on 9 June 1824 a meeting was held at the London Tavern, Bishopsgate Street,[2] to promote a ship canal, and in 1825 an Act was passed, giving the company a capital of £1¾ million.

The survey had been made in 1824. Telford took responsibility for it, but the work seems to have been done by James Green, and he it was who signed the three plans that were deposited[3] in connection with it. The proposal was for a large but orthodox canal to take ships of 200 tons; 95 ft wide at top, 15 ft deep, forty-four miles long, with thirty locks, and basins and harbours at Beer and at Stolford at the mouth of the Parrett on the north coast. A branch to Taunton was later added. The slump of 1825 and onwards put an end to what was always an unlikely scheme.

Meanwhile in Cornwall an old plan dating back to 1777 had been revived when a committee was formed in 1823 to consider the best means of forming a communication between Looe and Liskeard. Green was called in and asked to report upon the possibilities of a canal, a railroad, and a turnpike road. He was rushed in his work, which he completed in a fortnight, relying partly upon levels taken earlier. He reported that all three alternatives were practicable, but in the case of a canal "the descent of the Valley is too rapid to admit of a Canal being formed on the common principle of Lockage", the reasons being that the expense would be too great, the time taken to pass the locks too long, and the water supply insufficient. He ended:

> "I am therefore of the opinion that the major part of the Ascent must be overcome by means of Inclined Planes, and that the Canal should be made for Boats of Four Tons Burthen, which are well calculated for the Trade likely to pass over it."[4]

The lower part, however, was to be larger and locked, so that Looe river barges could reach their accustomed unloading place at Sandplace.

At this time the Liskeard and Looe was being thought of as an agricultural and coal-carrying canal, for it was only later that copper and other ores became important to it. Green's advice was not taken, and after an Act had been obtained in 1825, the canal was built with orthodox locks by a local engineer. It proved very successful.

Green was, however, consoled by a commission from one of the noblemen whose patronage he had solicited in 1808: John, Lord Rolle, one of the greatest landowners in North Devon. He was the son of Denys Rolle, who at the time of the canal mania had taken a great interest in tub-boat canals which would avoid the fertile river valleys and instead carry sea-sand to the hills on whose slopes it was needed. He had indeed dug an

[1] *Grand Ship Canal from Bridgwater to Seaton.* [2] *Gentleman's Magazine*, 1824, part 1, p. 633.

[3] Now in the Somerset County Record Office, Taunton.

[4] *Report of James Green, Civil Engineer, on a Canal, Rail Road, and Turnpike Road, between the Ports of East and West Looe, and the Town of Liskeard, in the County of Cornwall*, 1823.

experimental section of canal himself, helped by his coachman and his bailiff, to find out what was a reasonable day's work for a navvy.[1] A project of his in 1793 for a canal from Bideford to Torrington had in the following year grown to a proposal for one from Bideford to Okehampton, to link with another from Okehampton to Exeter. Then came the war; Denys Rolle died in 1797, and the idea was laid aside. In 1810, however, the original proposal for a canal from Bideford to Torrington was revived, and in August James Green is recorded as having produced a plan,[2] the result of what was probably his first canal survey in the west. It was not till 1824 that work was begun, without parliamentary authority and at Lord Rolle's expense.

The canal began at a sea-lock and basin communicating with the river Torridge below Wear Giffard, about one and three-quarter miles above Bideford Bridge. After a stretch of about a mile beside the river, the canal rose by a single inclined plane to its summit level. It followed the curve of the river to Beam, where a fine classical stone aqueduct of five round-headed arches carried it over the Torridge. The other bank was then followed to the warehouse, offices, and then existing Canal Tavern at the bottom of Mill Street, Torrington, and on to its end at New Manor Mill, also built by Green, a mile above the town. The purpose of the canal was principally to carry limestone. This was brought from South Wales to Appledore or Bideford, where it was unloaded into barges to be taken to the canal sea-lock. Here it was again transhipped to trains of tub-boats to be carried to Lord Rolle's kilns at Torrington. The first stone of the aqueduct was laid on 11 August 1824, and the canal was opened during 1825.

I have found no description of the plane, other than Green's own statement that it was worked by water. The aqueduct is a very fine piece of work, and would probably be more famous if the ordinary passer-by realized its original purpose. It is now used to carry a roadway over the Torridge to Beam House. New Manor Mill, a gloomy erection in a curious crenellated style that presumably appealed to Lord Rolle's military tastes, is also worth a visit.

We now come to the work for which Green is perhaps best known, the lifts on the Grand Western Canal. This canal had been projected during the mania and authorized in 1796, as a large barge canal. It was to run from Topsham on the Exe to Taunton to connect with the River Tone Navigation or with a possible Bristol-Taunton Canal; branches to Tiverton and Cullompton were also authorized. It was not started till 1810, and by 1814 a short length of the main line, from Holcombe Rogus west, together with the branch to Tiverton, had been built by John Rennie at a cost of £244,505,[3] rather more than the original estimate for the whole canal. Then work stopped because of the discouragement of the proprietors, and the short completed length carried limestone

[1] MS. letter from Denys Rolle in Salisbury and Southampton Canal records.

[2] *Exeter Flying Post*, 30 Aug. 1810.

[3] MS. account put in evidence, 1864. The information about this canal is taken from the records, mainly in the Tiverton Memorial Library and my own collection.

from Canonsleigh near Holcombe to the kilns at Tiverton, and coal brought to the end of the canal by land carriage from Taunton to be put into boats for the eleven miles to Tiverton.

If anything were to be done with the canal, it was essential to join it either to the Exe or the Tone. Of these, the junction with the Tone was the easier, and one or two suggestions had been made for accomplishing it before James Green approached the proprietors in 1829, his task having been made easier by the completion of the Bridgwater and Taunton Canal in 1827, which superseded the less efficient Tone Navigation, and promised water connection with a small but serviceable port for the import of coal.

A meeting was called, and James Green laid a report before it.[1] He began by showing that the charges for land carriage were so low that few traders thought it worth while to put coal into boats at Holcombe when it could be carried the whole way by land, and went on to say that little income would be got from the canal till "Coal, Culm and other articles can be conveyed from Taunton to Tiverton at a price much below the present rate of land carriage and this can only be accomplished by forming some mode of communication between the Canal near Loudwell Mill[2] and the Bridgewater and Taunton Canal near Taunton."

He said that a railway had been suggested, but that apart from the inconvenience of a double transhipment of cargoes, it would have to be very roundabout, with expensive cuttings and embankments. Otherwise there would have to be various steep ascents, with stationary steam engines. If a canal were chosen, it need not be on the original scale, but only on that of the Bridgwater and Taunton. Such a waterway would cost £100,000, and there was the objection that as the main traffic would be upwards from Taunton, there would be much waste of water from the main supply at Loudwell on the existing summit.

Instead, he suggested a small canal of 23 ft top width and 3 ft deep, with two or at the most three inclined planes worked by water. He pointed out that he had built the Bude and Torrington Canals with similar planes, and "they have been found to answer perfectly well", a statement that the harassed Bude Canal resident engineer, struggling with endless breakdowns on the planes, would not altogether have supported. If there were to be too much traffic for the water available, the inclines could be worked by steam engines. The cost would be £50,000, perhaps less, and Green ended by saying incautiously "I will if required engage to find responsible persons who will undertake the work with adequate security for that sum", and by throwing out the suggestion that if his proposal were successful a similar extension to join the Exeter Canal could be made on the same principles. A General Assembly was held in June; it agreed to Green's plan provided that the cost did not exceed £50,000.

At some time between April 1829 and April 1830 Green decided that it would be

[1] 28 Apr. 1829. [2] Near Holcombe Rogus.

better to build lifts than inclines. Canal lifts so far had been experiments, and none had been proved successful. The principle was obvious enough. Instead of hauling a small boat up an inclined plane, either on its own wheels or in a cradle or caisson, it should be possible to lift the boat vertically from one level to another. Either two caissons could be balanced on each side of a central set of wheels, or a single caisson could be counterbalanced by an equivalent dead weight. In either case a small motive power would then operate the lift.

Many had had the idea. Green himself says:

> "The merit of the first idea of passing boats from one pond of canal to another on this principle, is justly due to the late Dr James Anderson, of Edinburgh, who published a paper on the subject in his Agricultural Survey of the County of Aberdeen, about the year 1796; but it will be seen on a perusal of that able paper, that the details by which the principle was to be carried out, were left much to the practical man."[1]

Before this, however, Earl Stanhope and Robert Fulton had been corresponding on the subject in some detail.

The first effort to build a lift was, I think, Robert Weldon's. He began to build one, with a 45-ft rise, about July 1796 on the Somersetshire Coal Canal. As I understand the description,[2] this lift consisted of a single wooden caisson large enough to hold a full-length narrow boat carrying twenty to thirty tons, rising and falling in a totally enclosed chamber which was full of water and in which the caisson was immersed. By November 1797, the caisson could be raised or lowered in the shaft, and in May 1798 it had a successful trial. The walls of the shaft bulged, however, "a great expense was incurred, but the machine was found too unwieldy for common use, and after some trial was given up".[3] A second was built on the abortive Dorset and Somerset Canal near Mells. This, with two caissons counterbalancing each other, had a 21-ft lift and took ten-ton tub-boats. It was tested in September and October 1800, and again in June 1802, and appears to have worked satisfactorily, but because the canal was never opened it had no opportunity of proving itself in practice.[4] It was built by Mr Fussell, who was the owner of the local ironworks at Wells. A third, with a 12-ft lift, and with a single caisson balanced against an equivalent weight of bricks on a timber platform, was constructed on the Worcester and Birmingham Canal at Tardebigge in 1809. It did not work satisfactorily, and was soon abandoned.[5]

There is no evidence that Green was aware of any of these three experiments, or of

[1] "Description of the Perpendicular Lifts for passing Boats from one Level of Canal to another, as erected on the Grand Western Canal": *Trans. Instn. Civil Engineers*, 1838, ii, 185.

[2] For description and illustration see J. Billingsley, *General View of the Agriculture of Somerset* (ed. 2, 1798), 317–19. [3] W. Phelps, *History of Somersetshire* (1839), i, 59.

[4] J. Priestley, *Historical Account of the Navigable Rivers, Canals, etc.* (1831), 200.

[5] Worcester and Birmingham Canal records (British Transport Commission archives). See also L. F. Vernon-Harcourt, *Rivers and Canals* (1896), 399.

the correspondence between Lord Stanhope and Fulton, though as this correspondence took place in connection with the projected Bude Canal, of which Green was some twenty-five years later to be the engineer, it is possible that the Lord Stanhope with whom he worked made available his father's correspondence to the engineer.

An abstract of a report by Green was circulated before the Special General Assembly called for 13 April 1830. This proposed seven lifts, varying in height from 16 ft to 46 ft, and one inclined plane of 81-ft rise at Wellsford. His estimate was now £61,324 6s. 7d., and this was accepted. So confident was the engineer that he seems to have agreed to complete the lifts for a fixed sum.

At this point we must look at another project. In this same year of 1830 some Chard men got in touch with Green to discuss a possible canal to Chard, and Green in turn approached the proprietors of the Bridgwater and Taunton Canal, but no action followed. In 1833, however, the latter company heard of the projected improvement of the River Parrett navigation to Langport, and of the possibility of extending this navigation by a canal to Ilminster and Chard. They therefore employed Green to survey for a canal from their own navigation near Taunton to Chard. This he did, depositing the plan on 30 November 1833.[1] In his proposals as shown on this plan, he developed the ideas he was putting into practice on the Grand Western. In the latter case his problem was merely to build a canal up rising ground with a total increase in height of 262 ft. For the Chard, however, he had a worse problem, for he had to cross three ranges of hills with valleys between, and finish at Chard on top of a fourth. A locked canal would have been impossible in anything like a direct line, and he therefore conceived the idea of using planes and lifts in conjunction with tunnels to give him five canal pounds only, each higher than the one before. He planned to take the canal off the Bridgwater and Taunton at Creech St Michael, and to carry it over the Tone and across the flood meadows on a high embankment to his first lift at Thornfalcon, 28-ft rise, then through a tunnel (Lillesdon) and over another embankment to Wrantage. Here there was a second lift, 31 ft, and a second tunnel, a mile long (Crimson Hill), leading to a fairly level run to Ilminster, where there was to be an inclined plane of 94-ft rise to a third and short tunnel (Ilminster). Lastly, there was a fourth pound to Chard Common, a second plane of 78-ft rise, and the final length of waterway into Chard itself.

The Act for the Chard Canal was obtained in 1834, it having by now probably been decided to have four planes, and work was begun in June 1835. I do not know whether Green was employed as engineer on the Chard Canal in its early days, for the only reference I have is to the *Taunton Courier* of 19 August 1835, where it is stated that the contracts have been let at prices below Green's estimates. But he did not build the planes, nor is he afterwards mentioned in connection with the canal, and it seems likely that his troubles with the Grand Western caused him to be dismissed also from active

[1] Bridgwater and Taunton Canal records, and the deposited plan.

work in connection with the Chard, though that canal was actually built on his original line and in his way, though with four planes.

The cutting of the Grand Western extension began about 1832, and in February 1835, the first part, from Taunton to Bradford, was opened.

> "Seven barges, laden with coals, attended by another barge in which were a band of musicians and several spectators, yesterday passed through the aqueduct that crosses the Kingston road, having been brought into that higher level by the novel and efficacious process of lifting."[1]

All did not go well, however, and in their report to the Annual Assembly of 25 June 1835 the Committee, which had previously blamed both itself and the contractors for the delay in completing the extension, now put the blame on the engineer.

> "... The true cause appears to rest in the novelty of the plan of the Lifts, and the want of foresight as to actual inconvenience which might arise in using them. Although in January, 1834, the Lift near Taunton was reported to be ready and to answer every purpose, Mr Green relying with too much confidence on theoretical principles, never subjected it to a full and fair trial, so that many practical difficulties were only gradually developed and detected, which had not in the first instance been either observed or anticipated: as other Lifts came into operation, several serious obstacles to their free and perfect action were discovered and various expedients applied to remedy them without effect, till at length a secondary Lock Chamber was applied, besides improved arrangements of the machinery, all of which required time for their execution. The Engineer having engaged to complete these Lifts for a certain sum, all these experiments, alterations and additions, have been made at his expense, but, at a serious injury to the Company, from delay, and from the continued cost of superintendence without additional revenue: and to Mr Green at a heavy pecuniary sacrifice: That all these impediments are surmounted in a satisfactory manner the Committee are willing to believe, an opinion confirmed to a certain point by the opening of the Trade, including five Lifts, as far as Wellington."

The next year did not see the canal open, for the biggest lift on the canal, that at Greenham with a rise of 46 ft, had partially collapsed owing to subsidence, while the inclined plane at Wellsford would not work at all. In these circumstances James Green ceased to be engineer to the company on 27 January 1836.[2] It was not until mid-1838 that the canal was opened, and from that date to 1864, when the extension was closed, the lifts appear to have worked satisfactorily.

A detailed description of Green's lifts, by himself, will be found in the *Transactions of the Institution of Civil Engineers*, already quoted. They were not all quite similar, and Greenham lift, the last to be perfected, incorporated the experience gained in making

[1] *Taunton Courier*, 25 Feb. 1835. [2] *Taunton Courier*, 3 Feb. 1836.

the others work. Each lift consisted of two chambers separated by a pier of masonry. Within each chamber moved a caisson, wooden but strengthened with iron, and having an iron bar running lengthwise above it to which the lifting chains were attached. Each caisson was big enough to hold one eight-ton boat. At each end of the caisson were iron vertically-moving gates, and there were corresponding gates, one for each caisson, at the ends of the upper and lower pounds of the canal. The caissons, when at the bottom of the chamber, rested on wooden cross beams.

Above the masonry of the lift was a cast-iron frame on twelve supports, which held the 22-ft shaft upon which were mounted three iron wheels 15 ft in diameter. These wheels carried the chains that supported the caissons. Gearing connected to the centre wheel enabled the caissons to be moved by hand if necessary. Normally, however, the lift was operated by adding two inches, or one ton, of water to the uppermost caisson, when it overbalanced the lower. A brake, consisting of a lever working on a brake wheel, then controlled the descent. Because the weight of the lifting chains would alter the equilibrium as the weight of those holding the descending caisson became greater and those holding the rising caisson less as the caisson moved, corresponding lengths of chain were fastened below the caissons to compensate for these changes. When a caisson reached the top of the lift, a hand winch moved a forcing bar which pressed the caisson closely against the gate shutting off the canal pound. Both gates were then lifted together, and the boat floated out. At the bottom, the descending caisson was forced by iron wedges against the bottom gates.

The practice of excluding water from beneath the descending caisson, so enabling it to fall till it was level with the canal pound, Green had to learn by experience. At the lifts other than Greenham the caissons at first descended directly into the lower pound, and consequently would not sink far enough for the levels to be equalized. A shallow lock had therefore to be built below each lift into which the boat from the descending caisson was floated, after which the lock was emptied and the boat moved into the lower pound. The time taken to transfer a boat at these lifts was eight and a half minutes, against three minutes at Greenham.

The mistake Green made at the Wellsford incline is inexplicable. He designed this plane on the same principle as the Hobbacott Down plane of the Bude Canal, to be operated by the power of buckets filled with water and descending in wells.[1] He was responsible for building the Hobbacott Down plane, and in spite of its many breakdowns he apparently felt confident that the Wellsford plane would work on the same principle. Now at Hobbacott Down fifteen tons of water in a bucket were needed to pull a four-ton boat, running on its own wheels, up the incline. At Wellsford he was to

[1] Both Hobbacott Down and Wellsford planes had a double line of tracks, but both had to have enough power to haul a boat up even if nothing was coming down. In the case of Wellsford the descending empty cradle gave a small counterbalancing weight to the loaded rising cradle, but this made little difference.

handle eight-ton boats, and in addition he abandoned wheeled boats in favour of cradles on which the boats would be carried. The cradle had also, therefore, to be hauled. To provide the motive power he seems to have thought that only a little over ten tons of water in the buckets was enough, and built the plane accordingly. He was wrong, and the plane failed to work.

After he had ceased to be engineer, and after a local man, James Easton, had looked at it, W. A. Provis was called in. He reported on 28 June 1836 that about eight tons of water were needed to move the machinery and the empty cradle, and that to pull a loaded boat up the incline would require buckets holding twenty-five tons of water. Hydraulic power was thereupon abandoned, and a steam engine installed to work the plane.

So, ingloriously for one who clearly had a powerful character and a far-reaching mind, Green's career as a canal engineer in the south-west ended. After 1836 he did no other work on canals, though he contributed much to the Newport Docks and to improvements at Bristol, and in 1844 was consulted upon the building of the South Devon Railway because of his specialized knowledge of the estuary of the Exe.[1] In 1841 he left his post in Devon; "on relinquishing it he was presented by the Court of Quarter Sessions with a very handsome acknowledgement of the valuable services he had rendered the county". His obituary ends: "It was only in 1843, too late in his career, that he settled in London, for owing to the active competition of younger men, he was not so extensively employed as he might have been." He died on 13 February 1849.

[1] Miscellaneous papers relating to the S. Devon Railway Bill, 1844. Exeter City Library.

9 *Standards of living among canal boat people, 1840–1939*

W. FREER

The men, women and children who formed the labour force of the canal carrying trade in Britain between 1840 and 1939 have usually been regarded, both by contemporaries and by modern observers, as an impoverished, degraded social group, living a peculiar and anachronistic way of life on the fringes of more normal, respectable society.[1] However, my research indicates that, despite the decline of canal carrying during this period, canal boating could not be considered a depressed occupation. As a way of life and work it carried with it some serious social disadvantages but low pay and mass unemployment were not usually among them.

The so-called 'golden age'[2] of canal transport came to an end around 1840. After that date the rapid expansion of railways led to the neglect and abandonment of some waterways; several of the large canal carriers sold their boats and either transferred their business to rail or went out of business altogether.[3] One should not, however, assume that the railways took away such a large share of the canals' trade that there was hardly enough work left to enable the boatmen who remained to scrape a living. The proportion of the nation's freight carried by canal declined drastically during the nineteenth century so that, by 1900, the railways were carrying ten times as much freight as the inland waterways.[4] However, in absolute terms the volume of freight carried by canal remained fairly steady until the eve of the First World War. On many canals tonnage continued to rise after 1838, in some cases for long after. The Ministry of Transport, in 1941, estimated that at their peak in the 1830s the canals of Britain were carrying a total of about 40 million tons of freight.[5] By the time of the Royal Commission on Canals in 1906 the figure stood at just over 43 million tons.[6] Not until the First World War was there a really steep and continuing downward trend in the volume of traffic carried by canal.[7]

Therefore, although there was no great expansion in the trade, in general terms, there was no great contraction either. Certainly, on some canals, the effects of railway competition were so keenly felt that the waterway fell into rapid decline and was, in some cases, abandoned. But on others there was as much work available for boatmen, indeed a good deal more, than there had ever been before. For example, tonnage on the Birmingham Canal Navigations rose steadily from just over 3 million tons in 1838 to just under 9 million in 1898. On the Aire and Calder Navigation it rose from just under 1·5 million tons to nearly 2·5 million tons over the same period. Tonnage on the Warwick and Birmingham Canal, the river Trent and the Grand Junction Canal also increased during this period.[8]

While the volume of freight remained fairly steady, the size of the labour force contracted. Census occupational tables indicate that between 1851 and 1901 the number of men employed on inland boats and barges decreased by about 8 per cent.[9] During and after the First World War the decline in manpower was much more severe, something in the region of 50 per cent. However, although men left the declining trade for better prospects elsewhere, and on some canals boatmen were forced out of business, there is no evidence that large numbers of boatmen were forced into redundancy. In fact at various times throughout the history of canal carrying there was a shortage of skilled boatmen. This was true during the last few decades of the nineteenth century, during both World Wars and during the late 1930s.[10] During the First World War it was reported to the Board of Trade Canal Control Committee that boatmen were in such short supply that men could command double their normal wages.[11] When one of the principal long-distance carriers of the 1930s had to abandon ambitious expansion plans, shortage of boat crews rather than shortage of traffic was the principal reason.[12]

Therefore, whilst it is true that freight was carried less and less profitably as rates and tolls were cut in an attempt to combat railway competition,[13] increased productivity, reductions in underemployment and the continuing demand from some manufacturers for this form of transport meant that in many cases boatmen's money wages could be maintained at a reasonably high level.

In the same way that not all canals were able to withstand the onset of railway competition, not all sections of the carrying trade were successful in retaining traffic. During the nineteenth century the carriage of long-distance merchandise traffic was undertaken by large fly-boat companies. They operated lightly laden express boats which travelled non-stop, night and day, with frequent changes of horses and a crew of four men working round the clock in shifts.[14] In the first half of the century the fly trade was the most lucrative sector of the carrying trade. The men employed in it were better paid than their colleagues in the slow trade and the fly-boat captains seem to have enjoyed a higher status than those in the slow boats.[15] Fly-boat traffic, however, was also the most vulnerable to railway competition. By 1850 most of the well known private companies such as Pickford's had ceased canal carrying, and many canal companies opened carrying departments themselves at this time in order to salvage what they could of the general merchandise trade.[16] By the beginning of the twentieth century fly-boat traffic had almost died out.[17]

The vast bulk of canal traffic was always made up of heavy low-grade cargoes, mainly coal and other raw materials travelling, on average, less than twenty miles.[18] This was known as the slow trade, made up of smaller companies or even single-owner boatmen travelling mainly by day, their boats pulled by a single horse, usually provided by the boatman himself. It was here, within the slow, heavy bulk cargo sector of the carrying trade that the continued demand for canal transport was to be found.

After the First World War the distinction between the fly trade and the slow trade tended to disappear with the advent of diesel power, which allowed all boats to travel at the same speed. Motor propulsion was slow to make any impact on canal boat life. There were some experiments with steam power in the nineteenth century and one major narrow-boat carrier, Fellows Morton & Clayton, used steam propulsion for its fly boats between London and Braunston in Northamptonshire. However, there was no widespread use of motorised boats until the late 1920s or 1930s, and in some places horses remained in service until the 1960s.

The methods of remunerating boatmen were so various and elaborate that it is difficult to get an accurate idea of the average level of money wages at any time throughout the period in question. Most boatmen were not paid a fixed weekly wage but according to the length and difficulty of the journey and the value and tonnage of the cargo. Almost no two trips ever attracted the same payment. Furthermore, the diversity of canals, boats and boatmen makes it difficult to generalise about the income of boatmen as a whole. The following discussion, therefore, focuses mainly on men working long-distance narrow boats.

Before the twentieth century there is no evidence that full sets of agreed trip and tonnage rates were ever written down; if they were, none seems to have survived. It is possible to find copies of such agreements dating from the early years of the present century.[19] However, even where they exist, canal-boat work was so extremely irregular, with stoppages of days or even weeks occurring owing to the vagaries of the weather or delays in loading and unloading, that it is difficult to work out an average weekly wage from such documents.

An additional difficulty is that no fixed payment was made to the captain and each member of the crew. Only the captain was employed by the company, and he was left to hire his own crew and fix his own terms and rate of pay with them. Account also has to be taken of certain running costs which had to be met out of the payment made to the captain. These might include the cost of finding and keeping a horse or horses and the cost of ropes.

To get any idea at all of the level of money wages among boatmen one has to rely on estimates given by contemporaries – boat owners, canal inspectors and even boatmen themselves.

For example, in 1841 Richard Heath, a carrier from Stourport, stated that he paid his fly-boat captains on average £9 per week.[20] Out of this the captain had to find three horses and a crew of three men and pay various other running expenses. He could expect to net, on average, 30*s* a week. Figures given by Philip Bouverie of the Grand Junction Canal Company and Joseph Baxendale of Pickford's Canal Carrying Company confirm Heath's statement.[21] Slow-boat captains were paid less. A figure between 15*s* and 30*s* per week, excluding the cost of the horse, was suggested in 1841 by Josiah Hayes, an agent of the Staffordshire and Worcester Canal.[22] It probably resulted in an average of 20*s* a week net. This represents a money wage considerably above that of agricultural labourers of the time, and in fact boatmen's money wages were about on a par with those of coal miners and cotton spinners, two of the best-paid sectors of the working class.

As tables 1 and 2 indicate, for most boatmen money and real wages rose in the first seven or eight decades of the nineteenth century. Evidence given by factory inspectors in 1875 shows that slow-boat captains were earning between 25*s* and 42*s* per week after expenses.[23] This figure excludes the cost of the horse but not the cost of hired labour where that of the family was not used. Fly-boat captains were receiving about £4 17*s* 6*d* gross, or about 37*s* 6*d* net. This compares very favourably with the pay of other manual workers. After this time the fly trade did less well, but in the slow trade, up to 1885, the rate at which real wages rose was above the national average.

In common with the general trend in money wages, this rise was interrupted about 1890 and boatmen experienced a downturn in money wages around this time. Between 1900 and 1914 boatmen's earnings seem to have increased by something in the region of 14 per cent, but rising retail prices meant that the real increase was only half as much.

Nevertheless, boatmen continued to receive money wages on a par with, or even in excess of, those paid to other workers in the old staple industries. In 1906 a Board of Trade inquiry into earnings found that canal boatmen on piece rates earned on average 29*s* per week net. Engineering workers at that time were earning 28*s*, coal miners 32*s* and agricultural labourers 18*s* per week.[24]

During the First World War most industrial workers enjoyed significant increases in money wages. At first canal labour was left far behind and there was much discontent in the industry.[25] After 1917, however, when inland waterways came under government control, wages rose in what one member of the Canal Association described as 'the most reckless fashion'.[26] By 1920

Table 1. *Indices of money and real wages in the canal carrying trade, 1840–1910*

	Money wages (1840=100)			*Real wages (1850=100)*	
	Date	*Boat captains*	*National average*	*Boat captains*	*National average*
Fly boats					
	1840	100	100	100	100
	1845		98		
	1850		100	100	100
	1855		117		95
	1860		115		103
	1865		126		117
	1870		133		118
	1875	125	154	111	135
	1880	125	147	117	134
	1885		149		148
	1890	103	163	113	166
Slow boats					
	1840	100	100		
	1845		98		
	1850	100	100	100	100
	1855		117		95
	1860		115		103
	1865		126		117
	1870		133		118
	1875	125	154	111	135
	1880		147		134
	1885	150	149	156	148
	1890	115	163	126	166
	1895		162		158
	1900		179		180
	1905	150			
	1910	175			

Sources. National average figures from B. R. Mitchell and P. Deane, *Abstract of British Historical Statistics* (1962), Wages and the standard of living, pp. 343–7. Figures for boatmen based on information given in: *Select Committee on Sunday Trading*, 1841, House of Lords Journal, vol. 73, appendix 2; PRO RAIL 855/156, Oxford Canal Company Miscellany, p. 58; PRO RAIL 1021/77, Canal Association, report dated 16 August 1859; *Report of the Commissioners appointed to enquire into the Working of the Factory and Workshops Act*, Parl. Papers, 1876, XXIX, vol. I, appendix C, and XXX, vol. II, qq. 11274–82; *Select Committee on Canal Boats Act (1977) Amendment Bill*, Parl. Papers. 1884, VIII, qq. 542–3, 1136; *Royal Commission on Labour* (Group B), Transport and Agriculture, vol. III, Parl. Papers, 1893–94, XXXIII, between qq. 13,643 and 17,246.

they stood, according to one government source, at 100 per cent above pre-war levels,[27] although the data in tables 1 and 2 suggest a gain in the region of 70 per cent. In any case, the increase enjoyed by boatmen was significantly lower than that in certain other industries, most of which experienced increases of over 160 per cent, and significantly below the increase in retail prices, resulting in a net loss.[28]

Table 2. *Indices of money and real wages of slow-boat captains in the canal carrying trade, 1905–35 (1914=100)*

	Money wages		*Real wages*	
Date	*Boat captains*	*National average*	*Boat captains*	*National average*
1905	86	89	93	97
1910		94		98
1914	100	100	100	100
1920	172			
1925	163	196	93	112
1930	228	191	145	122
1935	163	185	114	130

Sources: National average figures from B. R. Mitchell and P. Deane, *Abstract of British Historical Statistics* (1962), Wages and the standard of living 1 and 2, pp. 343–7. Figures for boatmen based on information given in: PRO ED 11/88, Report of the Medical Officer of Health, Birmingham, 1906; *Board of Trade Report of an Enquiry into Earnings and Hours of Workpeople in the United Kingdom in 1906*, Parl. Papers, 1912–13, CVIII; PRO MT 52/14, Vernon Harcourt Lectures; PRO LAB 83/2480, 2492–6; *Committee of Enquiry into the Practice of Living-in on Canal Boats*, 1920–21; PRO HLG/52/135, Canal Boat Inspector's Report; PRO MT/52/42, Review by Minister of Transport, 1938.

After the war an initial boom was quickly followed by wage cuts in the early 1920s. In 1921 an industrial court ruling removed the wage increases granted to Midland boatmen in 1920[29] and a 5 per cent decrease was imposed on Fellows Morton & Clayton boatmen in 1923.[30] However, retail prices were falling so sharply at the time that real wage levels were maintained and even increased between 1921 and 1923. Money wages reached a peak around 1930. Falling retail prices again helped to augment this gain.

Between the mid-1930s and the outbreak of the Second World War the canal carrying industry experienced something of an upturn.[31] The reasons are not altogether clear. Ever since the report of the Royal Commission on Canals in 1906 had recommended the modernisation of the inland waterway system there had been pressure from some commercial interests for greater use of canal transport.[32] Their efforts met with very little success but it seems that during the 1920s, possibly as a result of the motorisation of narrow boats, there was a burst of optimism about the future of canal transport. The Grand Union Canal Company was formed and a programme of improvements to its main line was embarked upon. This was quickly followed by the formation of the Grand Union Canal Carrying Company, a particularly innovative concern. It envisaged a bright future for a modernised industry which it did its best to bring about.[33] It began a major programme of boatbuilding, set up subsidiary companies to offer a fuller service to customers and, for a while, succeeded in attracting much new traffic to the canal. The other major narrow-boat carriers of the period, Fellows Morton & Clayton, were obliged to follow suit as far as possible and they too began renewing and expanding their fleet.

All this activity and optimism, however, had little effect on boatmen's earnings, which fell during the decade by something like 30 per cent. Nevertheless, money wages continued, on the face of it, to compare favourably with those of some other industrial workers. For example, in 1936 narrow-boat captains on the upper Trent were receiving on average just over 60*s* per week net.[34] In 1935 engineering workers were averaging 55*s* a week, coal miners 45*s* and agricultural workers 35*s* weekly.[35] However, the boatmen worked a considerably longer week than the average industrial worker's 48·9 hours. Boats were said to work twelve hours a day or more, and a seven-day week was common. Unavoidable delays meant that the pace was by no means relentless but the irregularity of working patterns meant that men were sometimes significantly worse-off than the average figures suggest.

Men on larger boats on the broad canals of the north could expect a more regular income, with the captain and other members of the crew receiving a fixed amount. Although men on larger vessels tended to be better paid than those on narrow boats, wage agreements concluded during the inter-war years show that such was not always the case. For example, in 1920 narrow-boat captains could net between £2 16*s* and £5 a week, whereas the captains of four-man motor boats in the north-west were earning £2 plus a tonnage rate which might mean a total of £3 a week. Throughout the 1920s and 1930s men employed on these larger boats, whether captain or crew, were paid significantly above the average money wages of workers in the old staple industries.[36]

In addition to these apparently high rates of money wages, those who lived permanently on board (admittedly a minority of boat people – see below) had no rent, rates, fuel or travelling expenses to find. Poaching from canalside fields, fishing from the canal itself and 'perks' from the cargo which might be used by the boat crew or exchanged at locks for other goods also enhanced the crew's income.

Given that documentary evidence on boatmen's money wages is scant and hard to interpret, it is necessary to cross-reference these findings with other indicators of living standards.

The idea that canal boating was an impoverished occupation seems to have arisen partly out of a mistaken belief that it was customary for boat families to live on board their boats. It is commonly supposed that the practice began around the middle of the nineteenth century, when poverty, brought on by railway competition, forced boatmen to give up their home on shore and replace their hired hands with the labour of their own families.[37] In fact, as the canal historian Harry Hanson has pointed out,[38] the testimony of several witnesses to at least two parliamentary select committees in the first half of the nineteenth century indicates that families were commonly found living and working on boats before 1830. John Hassell, describing a journey on the Grand Junction Canal undertaken in 1819, commented that women were commonly to be found working on canal boats at that time, long before the advent of railway competition.[39]

There was clearly some tendency for living in on boats to increase during times of poor trade. For example, according to the census, between 1851 and 1861 there was an increase of approximately 30 per cent in the number of boatmen living on board with their families, and a small increase during the course of the following decade, but thereafter the proportion of boatmen living on board with their family began to decrease. This trend continued until the First World War. The practice of living on board was more prevalent in some regions than in others.[40] Fewer families lived on board in the north of England. On the route between the Severn ports and Birmingham, significantly a narrow canal route, it was more usual for boats to be worked by a man and a boy, sometimes a hired hand or the captain's son.[41] Even when Worcester and Gloucester boat families did sleep on board, most had a house on the bank in Gloucester or Worcester and were not permanently resident on board. This is evident from the testimony of various witnesses, canal inspectors and company managers, even schoolteachers, and by the census enumerators' returns which show that the children of Gloucester and Worcester boatmen all tended to be born in the same place, unlike those of some other boatmen whose children were born in various canalside towns and districts.[42]

On the Grand Junction Canal and on canals in Staffordshire and Warwickshire it was far more common to find whole families living on board, regardless of the state of the trade. Hence, the use of family labour on canal boats seems to have been a matter of established regional custom as well as economic necessity.

Most important of all, the proportion of all boatmen living on board and relying on family labour was always quite small. Most canal journeys were short, undertaken by boats with no living accommodation. Fly boats never had families living on board. Boatmen in the fly trade, if they were married, maintained a shore-based home. Census data suggest that in 1851 no more than 18 per cent of all boatmen were living on board with their families, 26 per cent in 1871 but no more than 21 per cent between then and the First World War.[43] Moreover, these percentages are the highest possible figures. In reality they may have been much lower.

What is even less generally appreciated is that, even among this minority of boat families who slept and worked on board, few lived there permanently. Most had a shore-based house in which all or part of the family lived from time to time. There are numerous witnesses to testify to this during the nineteenth and early twentieth centuries, not only boat owners but factory inspectors, canal inspectors, schoolteachers and missionaries. For example, a survey conducted by the Canal Association in 1884 found that, of families living on board in May that year, 70 per cent had houses on the bank.[44] Another survey conducted by a government inquiry in 1920 found that less than 50 per cent of families living on board had no shore-based home[45] and at other times, and in particular regions, the percentage was much lower.

The myth that poverty, whether induced by railway competition or not, forced boatmen to give up maintaining their families ashore and to take them to work on the boats[46] seems, in part at least, to have been created and maintained by some boat owners and by a section of the boating population itself. This is particularly noticeable from the 1870s onwards, when the argument that a canal boat was no place in which to bring up a family[47] led to attempts[48] to prohibit children from living on board. Any such suggestion was always met with an outcry from those boatowners and boatmen who relied on family labour, claiming that they could not afford to do otherwise. The real motive seems to have been partly a genuine desire to keep boat families together and perpetuate their traditional way of life. It was also commonly believed by people associated with the canals that children not brought up on the boats would never take to the work in later life and that the source of canal boat labour would thus dry up. There was in addition a strong element of the boatman wanting to enhance his net income by substituting the labour of his wife and children for that of a (male) hired hand. Boatmen who used family labour claimed that they could not afford to hire hands instead.[49] That this was not the whole truth is evident from the fact that many boatmen managed to carry on without family labour.[50] It is also possible that had the family remained at home its members might have been able to find alternative employment, although whether any such additional income would be sufficient to offset the cost of hiring a male hand would obviously depend on individual circumstances.

The study of kinship patterns among boat people can also be used to indicate levels of prosperity within this community. The following remarks are drawn from a detailed study, based on census enumerators' returns, of people living on board boats in Birmingham, Shardlow, Braunston, Brentford and Paddington, and also of boat families living in houses in the rural community of Braunston between 1841 and 1881.[51]

In his study of working-class kinship structures in Lancashire[52] Michael Anderson found that the industrialising period saw a widening of kinship patterns. He discerned an increasing tendency for households to contain persons other than the nuclear family of the head of the household. This he attributed to growing working-class affluence, which allowed heads of families to accommodate people who were not their immediate dependants. The tendency can also be seen among the land-based boat families in Braunston. The percentage of families containing persons other than members of the current nuclear family of the head of the household rose from just over 17 per cent in 1851 to 36 per cent in 1881.[53] According to Anderson, 23 per cent could be considered a high proportion even in comparatively highly kinship-oriented communities.[54]

The widening kinship patterns in Braunston were to some extent a survival strategy adopted in response to declining trade, but the ability to extend aid to kin outside the immediate nuclear family does indicate a certain degree of relative affluence.

Most of the non-nuclear family groups suggest a high degree of flexibility. For example, in Braunston in 1851[55] two children, William and Sarah Howell, aged nine and five respectively, were living with their maternal grandmother, Ann Warwick. In 1861[56] their father, John Howell, was staying with his mother-in-law, the said Ann Warwick, but the children, teenagers by then, had either died or left to live elsewhere. A married son of Ann Warwick and his young family were also living in the house but there was no sign of Howell's wife or any other children belonging to him. On the night of the 1871 census Howell, his wife and a seventeen-year-old unmarried daughter, Eliza, were sleeping on a boat on the Regent's Canal.[57] Ann Warwick remained in the village but her married son had moved with his wife and seven children to a separate house, leaving only his fifteen-year-old daughter with Ann.[58] In 1881[59] Ann Warwick, probably deceased, was no longer present and John Howell was back in Braunston with his wife and two unmarried daughters, living as a single nuclear family. One daughter was the previously mentioned Eliza, now said to be aged twenty-four, but the other, Louisa, aged twenty, appeared for the first time. There had been no indication in the previous census of where she had been living even though she would have been a child of ten at the time.

Despite this high degree of flexibility displayed by some extended family units, a highly functional orientation towards kinship seems to have been the norm.

Most Braunston boat families had a home on the boat and in the village at one time or another. Some families maintained two homes simultaneously, others locked up the village home[60] when they went away on the boat or, for some periods of time, had no shore-based home at all. Where a family had too many children to be accommodated on the boat, or it became desirable to leave some children for other reasons, a shore-based home became necessary. It might be needed at other times, too, when the boat, for some reason, was unavailable or unsuitable. If the family had no home of its own in the village, members of the extended family living ashore could be called upon to offer this facility.

Reciprocity seems to have occurred rarely, suggesting a high degree of trust over a fairly long time span. The significance of this as an indicator of living standards is that the willingness to offer aid, even when reciprocity might be remote, indicates a certain amount of surplus resources.

The census enumerators' returns can be used to deduce information about marriage, childbirth and the residence of young couples which is another indicator of relative prosperity. Braunston-based boatmen and their wives tended to be quite young at the birth of the first child (in 1841, the average age of the wife was twenty-one, rising to twenty-four by 1881),[61] which indicates early marriage as normal for men and women. The average figures conceal the fact that it was not uncommon for girls to give birth to their first child in their late teens.[62] In most cases the husband was between two and five years older than his wife and in a few cases he was actually younger.

Unlike other members of the community, a young boating couple, if they were willing to live on the boat, did not have to worry about living with parents or in-laws or affording a home of their own. This can be seen in the Braunston community, where very few young boat couples can be traced living with relatives after marriage. Even when the wife was left behind in the village while the husband travelled with the boat, most young couples lived in a home of their own before the birth of the first child. This seems to indicate the relative prosperity of young boatmen.

There are some aspects of these changing kinship patterns which indicate worsening economic circumstances rather than the reverse, with members of the immediate family replacing a paid crew and extended aid networks developing to offer facilities to families who could no longer afford to maintain a home on shore as well as on the boat. However, although times were getting harder in the canal carrying trade, they had not reached the point where resources had become so scarce that no aid could be extended to relatives outside the current nuclear family unless an immediate return was expected.

If the general standard of health in a community may be taken as an indication of living standards, here again it is possible to point to evidence which supports the view that boat people enjoyed a standard of living which was, in some respects, above average. Although Hanson claims that the canal boat population was almost universally regarded as one of the healthiest in England,[63] there is no real evidence that anyone took the slightest notice of the health of boat people until the last quarter of the nineteenth century. But, from about 1885 onwards canal boat inspectors and, later, school medical officers and Medical Officers of Health generally spoke favourably of boat people's health. A Ministry of Health committee of inquiry into the practice of living-in on canal boats in 1920 found that in matters of health, cleanliness, morality, feeding and clothing boat people were the equals, if not the superiors, of town dwellers *of a similar class*. The qualification, often echoed even more bluntly in the comments of other observers, indicates, however, that the standards applied were not high.

Mortality statistics published by the Registrar General show that in 1860 the death rate of boatmen, bargemen and watermen was 22·8 per thousand, only 0·7 per thousand higher than that for all males. During the next forty years it fell, but at a slightly slower rate than that of the general male population. Analysis of the causes of death shows that in 1890 boatmen were nearly four times as likely as other males to be killed in an accident.[64] This cause of death apart, the statistics would therefore suggest a standard of health at least as good as average.

Death from alcoholism and circulatory disease was also significantly higher among boatmen than among the general population.[65] Although alcohol abuse has never been confined to the better-off, and heart disease can be attributed to causes other than fatty diet, both causes of death do seem to say something about consumption among boat people which indicates a certain level of surplus resources. That boatmen 'ate well', that is to say, consumed a large amount of red meat, butter and cheese, is confirmed by contemporary witnesses from the late nineteenth century onwards.[66] Furthermore, in the 1930 mortality statistics, death from digestive disease was common among boat people. The death rate from most diseases increases as one goes down the social scale. With digestive disease, in 1930 at any rate,[67] the reverse was true. Whereas death rates from other complaints put boatmen in social class 4 or 5 (5 being the lowest), boatmen were as likely to die of digestive disease as men in social group 1. Although it is not clear what exactly constituted digestive disease, if it is linked with diet then it would be a further indication that boat people enjoyed what would have been considered at the time a good diet, i.e. one containing plenty of meat and animal fat. This seems to indicate a high level of spending on food.

There is room to comment only briefly on standards of housing and education. As already indicated, the great majority of boat people did have a house on shore, although many spent little time in it. The type of housing inhabited by the boating population in the nineteenth century was probably similar to that of the unskilled and semi-skilled labouring classes in rural districts.

For those who did live on board, accommodation compared favourably with average early to mid-nineteenth-century working-class housing in most regards except space. The cabin provided extremely cramped quarters for a family, although overcrowding was no worse, and probably better, than among the section of the population housed in one or two-room tenements in the second half of the nineteenth century. From the beginning of the twentieth century, however, standards of accommodation on board compared ever less favourably with those on shore and by the inter-war years they were far from satisfactory.

Furthermore, there is no doubt that those who lived mainly or wholly on board were poorly served with amenities such as schools, medical facilities and shops. Voluntary agencies such as the London City Mission and the Incorporated Seaman and Boatmen's Friend Society tried to meet the peculiar needs of the way of life of those living on board but provision fell far short of needs. Education was a particularly intractable problem. School registers, the reports of school attendance inspectors and NSPCC inspectors, boat children's attendance books, government surveys, even the boat people themselves admitted that where children lived wholly or mainly on board they attended school irregularly, if at all.[68] In some areas, even when children lived ashore, they were often taken away on the boat from about the age of eight, and school saw them no more.[69] Even as late as 1939 many boat children received no education worth mentioning, and child labour remained a common abuse throughout the history of canal carrying.

The nature of the sources, the complexity and irregularity of canal boat work and its remuneration and the diversity of canals and boatmen mean that these findings must remain tentative and should be treated with caution. Such figures as exist for boatmen's earnings in the nineteenth century are few and far between. Rather than giving a substantial data bank on which to base a reliable series of real or money wages, they represent isolated glimpses at different times and in different places. Furthermore, some rely for their accuracy on the honesty of men who may at times have felt it to be in their interest to represent the boatman as respectable and relatively prosperous.[70] By about 1900 the statistics start to become more reliable, although only slightly more plentiful, and figures given in official surveys of the Board of Trade and in wage agreements between trade unions and employers broadly support estimates reached by other observers. There remains, however, the diversity of canal boat work, which makes it difficult and dangerous to give broad averages. This article has concentrated mainly on the living standards of men working long-distance narrow boats, but there were many other men working the canals whose circumstances may have been quite different.

Conclusions about kinship patterns based on census data suffer from the usual complaint that the census provides only a decennial snapshot of the community in question, and the interpretation of the relationship between health statistics and standards of living is far from straightforward. Seldom do the sources allow the boat people to speak for themselves. When they do, the main intention seems often to have been to say whatever was most likely to preserve a traditional way of life.

Despite these reservations, there seems to be enough broad agreement between the different sources to conclude that, throughout most of the history of canal carrying, many long-distance boatmen received money wages which compared very favourably with those of other manual workers. Where boatmen chose to take their families on board, it was often a matter of regional custom or personal choice rather than because they could not afford to do otherwise. However, when a boat family did decide to live wholly or mainly on board, from the late nineteenth century onwards, they did so at considerable social cost. Cramped, insanitary, primitive accommodation on boats, illiteracy and ignorance gave outsiders the impression that boat people were living in poverty and misery. In some cases it may have been true. For many, however, the determination to preserve the family unit and a preference for the peculiarly independent way of life of canal boating as against the tyranny of the factory meant imposing upon themselves standards of living significantly below what they might have afforded ashore. It is interesting to note that social deprivation and some forms of poverty are not merely a reflection of low pay.

Notes

1 See, for example, *Select Committee to enquire into the Expediency of restraining the Practice of carrying Goods and Merchandise on Canals and Navigable rivers and Railways on Sundays, House of Lords Journal*, 4 and 5 Vic. (1841), vol. 73, appendix 2, qq. 84, 109; *Report and Evidence of the Commissioners appointed to enquire into the Working of the Factory and Workshops Act*, Parl. Papers, 1876, XXX, q. 10,500; George Smith, *Our Canal Population: the Sad Condition of Women and Children – with Remedy* (1875); Peter L. Smith, *Canal Barges and Narrow Boats* (Aylesbury, 1975).

2 This is generally agreed by canal historians. See, for example, Charles Hadfield, *British Canals: an Illustrated History* (Newton Abbot, 1950), and Harry Hanson, *The Canal Boatmen, 1760–1914* (Manchester, 1975). However, for the idea that canal transport at its peak did not enjoy such an unchallenged position as the term 'golden age' seems to suggest see Gerard Turnbull, 'Pickford's and the canal carrying trade, 1780–1850', *Transport History*, VI (1973), p. 20.

3 Turnbull, 'Pickford's and the canal carrying trade', pp. 24–5; Alan H. Faulkner, *The Grand Junction Canal* (Newton Abbot, 1972), p. 170.

4 D. L. Munby, *Inland Transport Statistics, Great Britain*, ed. A. H. Watson, 1, table A10, p. 83. The table gives a figure of 418,096,000 tons of freight (excluding freehauled) carried by rail in 1900.

5 PRO MT 52/109, Report of Frank Pick to the Ministry of War Transport, May 1941.

6 *Royal Commission on Canals and Inland Navigations in the United Kingdom*, Parl. Papers, 1906, XXXII, I, Part II, appendix 1, statement 1.

7 PRO MT 52/4, Canal Control Committee, 5 February 1919.

8 *Royal Commission on Canals*, 1906, I, Part II, appendix 1, statement 1.

9 *Census of Great Britain*: 1851, *Population Tables*, II, Part 1, table XXV, p. ccxxiii; 1861, *Population Tables*, summary table XVII, p. xxxi; 1871, *Population Abstracts*, III, table XVIII, p. xxxviii; 1881, *Population Tables*, III, table V, p. xi; 1891, *Population Tables*, III, table V, pp. xii–xiii; 1901, *General Report and Appendices*, table 34, p. 259; 1911, *General Report with Tables*, table 1, p. 17; 1921, *Occupation Tables*, table 1, p. 11; 1931, *Occupation Tables*, table 1, p. 14; 1951, *Occupation Tables*, table 1, p. 14; 1961, *Occupation Tables*, table 1, p. 4.

10 *Factory and Workshops Commission*, PP, 1876, XXX, q. 10513, PRO MT 52/1, Minutes of the Board of Trade Canal Control Committee, 28 May 1917; PRO MT 52/42, Review by the Minister of Transport, Report of a meeting held on 19 January 1938 between the Minister and the manager of the Grand Union Canal Carrying Company; PRO

MT/43, Minutes of the Ministry of Transport Canals (Defence) Advisory Committee, 28 November 1939; PRO MT 52/109, Report of Frank Pick to the Ministry of Transport, 1941, para 48.

11 PRO MT 52/1, Board of Trade Canal Control Committee, 28 May 1917.

12 PRO MT 52/42, Review by Minister of Transport, minutes of a meeting held on 19 January 1938.

13 Hadfield, *British Canals* (1950 edition), p. 181.

14 *SC on Sunday Trading, House of Lords Journal*, 73, 1841, qq. 532–2.

15 *Ibid.*, qq. 217, 1221.

16 Turnbull, 'Pickford's', pp. 24–5; Faulkner, *Grand Junction*, p. 170.

17 *Minutes of Evidence, Select Committee on Canals*, Parl. Papers, 1883, XIII, qq. 2666–7.

18 Gerard Turnbull, 'Canals, coal and regional growth during the industrial revolution', *Economic History Review*, XL (1987), p. 543 n. 21.

19 PRO LAB 83, pieces 2382–509.

20 *SC on Sunday Trading, House of Lords Journal*, 73, 1841, qq. 626–9.

21 *Ibid.*, qq. 1053–163.

22 *Ibid.*, qq. 216–20.

23 *Factory and Workshops Commission*, PP, 1876, XXX, qq. 11274–82.

24 *Board of Trade, Report of an Enquiry into the Earnings and Hours of Work People in the United Kingdom in 1906*, Parl. Papers, 1912–13, CVIII.

25 PRO MT 52/1, Minutes of the Canal Control Committee, 1 June 1917, refers to and quotes from correspondence placed before the committee with regard to the widespread discontent in the industry. PRO LAB 34/29–38, Strikes and lock-outs, 1911–20.

26 PRO MT 52/5, Transcript of a conference between the Ministry of Transport and representatives of the Canal Association, 22 July 1920.

27 PRO MT 52/14, R. B. Dunwoody, *Vernon Harcourt Lecture*, 1920, p. 40.

28 B. R. Mitchell and P. Deane, *Abstract of British Historical Statistics* (1962), Labour force, table 26.

29 PRO LAB 83/2494, Industrial Court Minutes, 27 April 1921.

30 PRO LAB 83/2496, Industrial Court Minutes, 20 November 1923.

31 PRO MT 52/109, Report of Frank Pick to the Ministry of Transport, 1941, para. 61; PRO MT 52/42, Report of a meeting held between the Ministry and the Grand Union Canal Carrying Company, 19 January 1938.

32 For example, see PRO MT 52/14, which contains reference to the Waterways Association comprised mainly of Midland traders and manufacturers and chaired by Neville Chamberlain. It came into existence between 1906 and 1914 to press for the implementation of the recommendations of the Royal Commission on Canals and Inland Navigations in the United Kingdom, 1906–09.

33 PRO MT 52/42, Review by the Minister of Transport, 1938; HLG 52/141, Ministry of Housing and Local Government, Canal Boats, General Inspectors' Office Committee; MT 52/123, Proposals of the Grand Union Canal Company, 1942–48.

34 PRO LAB 83/2492, Award in the matter of arbitration between the Trent Navigation Company Ltd and the Transport and General Workers Union, 20 July 1936.

35 John Benson, *The Working Class in Britain, 1850–1936* (1989), p. 41.

36 See list of sources at the foot of table 2.

37 See, for example, Hadfield, *British Canals* (1966 edition), p. 73; *The Canal Age* (1968), p. 151; Smith, *Canal Barges and Narrow Boats*, p. 15.

38 Harry Hanson, *The Canal Boatmen, 1766–1914* (Gloucester, 1984), pp. 53–5.

39 John Hassall, *A Tour of the Grand Junction Canal in 1819*, (reprinted 1968), p. 45.

40 *Census of England and Wales*, population and occupational tables in the county reports for the years 1851–1971. For a more detailed analysis of these data see Wendy Freer, 'Canal Boat People, 1840–1970', unpublished Ph.D. thesis (Nottingham, 1991), pp. 133–47.

41 *Census of England and Wales*, enumerators' books for boats tied up at Worcester Bar, Birmingham, on the Worcester and Birmingham Canal; 1861, RG 9 2139; 1871, RG 10 3089, 3100; 1881, RG 11 2877, 2988.

42 See for example PRO HLG 52/141, report by Dr Stock of the Ministry of Health to Gloucester, 1934; *Report and Evidence taken before the Committee appointed by the Ministry of Health to Inquire into the Practice of Living-in on Canal Boats in England and Wales and to Report whether any Alteration in the Practice is Desirable*, 1920–21, q. 1288, evidence of Alfred Colwell, schoolteacher, Gloucester, and q. 1666, evidence of E. Manning-Lewis, general manager of Sharpness New Docks and Gloucester & Birmingham Navigation Company. (This is an unpublished report, copies of which can be found among the papers of the Ministry of Housing and Local Government and the Board of Education at the Public Record Office, for example ED 11/88.)

43 For a more detailed analysis of these statistics see Freer, 'Canal Boat People'.

44 *Select Committee on the Canal Boats Act (1877) Amendment Bill*, Parl. Papers, 1884, VIII, appendix 1.

45 PRO ED 11/81, statistics prepared for the Committee on Living-in, 1920–21.

46 Hadfield, *British Canals*, p. 73; *Canal Age*, p. 151; Smith *Canal Barges*, p. 15.

47 A campaign based on this argument was started in the 1870s by George Smith of Coalville, author of *Our Canal Population*, whose agitation led to the passing of the Canal Boat Acts, 1877 and 1884. For further biographical details see Edwin Hodder, *George Smith of Coalville, the Story of an Enthusiast* (1896).

48 Examples of such attempts are *The Committee on Living-in*, 1920–21, and a Private Member's Bill to prohibit children from residing on canal boats, introduced by Harry Gosling MP 1929–30. Details of this Bill can be found at PRO ED 24/1122, ED 31/263, HLG 52/135–7.

49 *Committee on Living-in*, 1920–21, qq. 489, 2252, 2271–3.

50 *Ibid.*, evidence of NSPCC inspector Hackett, qq. 1548–55, *Factory and Workshops Commission*, PP, 1876, XXX, evidence of canal manager of North Staffordshire Railway, qq. 1548–55 and evidence of two boatmen, qq. 10868–916.

51 Census enumerators' books used in this study can be found under the following references. Brentford: 1841, HO 107 655; 1851, HO 107 1698–1699; 1861, RG 9 777; 1871, RG 10 1319; 1881, RG 11 1349. Birmingham: 1841, HO 107 1147; 1851, HO 107 2052, 2057; 1861, RG 9 2138, 2139; 1871, RG 10 3089, 3100; 1881, RG 11 2877, 2988. Braunston: 1841, HO 107 798; 1851, HO 107 1741; 1861, RG 9 945; 1871, RG 10 1492; 1881, RG 11 1559. Paddington: 1861, RG 9 2, 3; 1871, RG 10 2, 8, 13, 25; 1881, RG 11 11, 4, 18.

52 Michael Anderson, *Family Structure in Nineteenth-century Lancashire* (Cambridge, 1971), p. 1.

53 Census enumerators' books HO 107 1741, RG 9 945, RG 10 1492, RG 11 1559.

54 Anderson, *Family Structure*, pp. 43–4.

55 Census enumerators' book HO 107 1741.

56 Census enumerators' book RG 9 945.

57 Census enumerators' book RG 10 9.

58 Census enumerators' book RG 10 1492.

59 Census enumerators' book RG 11 1559.

60 Census enumerators' book RG 11 1559. In this census, the enumerator actually notes fifteen houses in Braunston which are empty owing to the occupants being away on the boats.

61 Census enumerators' books HO 107 798, 1741, RG 9 945, RG 10 1492, RG 11 1559.

62 For example, Census enumerators book HO 107 798, Hannah Hopkins, aged nineteen, child aged four months, and census enumerators' book RG 9 945, Mary Jones aged twenty-nine, eldest child aged ten years.

63 Hanson, *Canal Boatmen*, p. 131.

64 *Supplement to the Sixty-fifth Annual Report of the Registrar General for England and Wales, Births, Deaths and Marriages, 1891–1900*, Part II, Parl. Papers, 1908, XVII.

65 *Supplement to the Fifty-fifth Annual Report of the Registrar General of Births, Deaths and Marriages in England*, Part II, Parl. Papers, 1897, XXI; *The Registrar General's Decennial Supplement, England and Wales*, 1931, Part IIA, Occupational Mortality, tables 4D, 6.

66 *Committee on Living-in*, 1920–21, q. 354 and appendix 1, para. 7.

67 *The Registrar General's Decennial Supplement, England and Wales*, 1931, Part IIA, Occupational Mortality, tables 4D, 6.

68 For a few examples of sources which support this see: PRO ED 11/1, attendance register of boat children at a school in Mirfield, June 1887; *Morning Leader*, 3 October 1900; PRO ED 11/40, Minutes of the Northwich Rural District Council for 14 February 1905; PRO ED 11/87, Report of the Association of Education Committees, 26 June 1914; *Times Educational Supplement*, 22 November 1917; *School Master*, 17 May 1919; PRO ED 11/88, letter from the National Transport Workers' Federation to the Board of Education and Ministry of Health, 28 August 1919; written and verbal evidence given to the Committee on Living-in, 1920–21; PRO ED 11/40, Education of Children on Canal Boats, Summaries of Annual Returns to the Board of Education, 1902–05; PRO ED 11/88, Canal Boat Children, Returns, 1917–21; *Lock and Quay*, house magazine of the Docks and Inland Waterways Executive, March 1952, article on the education of boat children, March 1959, article on the special canal boat school at Norwood, Southall, May 1960, interview with boatman, George Beechey.

69 PRO ED 11/88, Archdeacon Street School, attendance register and remarks of teacher about boat boys attending the school, 3 November 1919. PRO ED 11/41, Report of the School attendance inspector, Sheffield, 1908.

70 For example, it could be argued that this was the case with some boatowners appearing before the Select Committee on Sunday Trading, *House of Lords Journal*, 73, 1841, who did not want their business stopped on Sundays on the grounds that their boatmen were degraded, immoral, etc., because they did not attend church.

10 *The canal boatmen's strike of 1923*

K. B. SHERWOOD

On 13 August 1923 the movement of long-distance commercial traffic on those canals linking London, the Midlands and north-west England came to a virtual standstill when the Transport and General Workers' Union (TGWU) called out on strike those of its members employed by Fellows Morton & Clayton (FMC), one of the country's largest canal carrying companies.[1] This dispute was to last for fourteen weeks, though in the context of industrial turmoil between 1918 and 1926 it has been an event barely recalled either in the economic and social histories covering this period or in the published recollections of those personnel most actively involved.[2] Nevertheless, despite this apparent insignificance, a narrative of the origins and course of the strike can be justified for three specific reasons. First, not only was it the sole strike in FMC's history (and, perhaps, the longest ever to involve any single canal carrying company), it was also one of the first disputes to involve the recently created TGWU. To analyse this dispute, therefore, is to trace the expansion of the union's sphere of activity into non-river-based trades and occupations, and also to examine the interaction between a company with no strike record, let alone negotiating experience, and a large, but young, national union keen to flex its industrial muscle on behalf of a neglected section of industrial society. Second, the dispute can be related to the particular characteristics of the canal carrying industry and to its changing role within the national freight distribution system. Thus, on the one hand, the strike was part of the union's campaign to improve the working conditions of canal boatmen, whilst on the other hand its causes can be traced to the competition between canals and railways for the carriage of bulk traffic, and to the growing rivalry of road transport to both. Third, in an industry where long-distance carrying was still dominated by family boats, the strike provides a unique opportunity to study a section of the canal community at close quarters. One outcome of the strike call was that FMC boats and their occupants gathered at those locations where the company had depots, and it was to these depots that the local and national press were drawn to comment on the background to the dispute and to describe events and activities. Most of the depots were in large towns and cities, and so the impact of the dispute upon the local community was slight: the exception was that at Braunston, a village near Daventry, Northamptonshire, at the junction of the Grand Junction (now Grand Union) and Oxford Canals. It is the purpose of this article, therefore, to examine the boatmen's strike of 1923 within the context of this small canalside settlement in north-west Northamptonshire.

The canal at Braunston

To most boatmen operating between the major urban centres of midland England, or between Birmingham and London, the name of Braunston was synonymous with the canal, for one consequence of the most active period of waterway construction, between 1770 and 1815, was the emergence of this village as the focal point of the canal system in the south Midlands. The strategic importance of Braunston had been recognised from the outset when, in 1788, the village was selected by Pickford's, then a canal-based carrying company, as one of the major depots in its national distribution system.[3] With the completion of the Grand Junction, Warwick & Napton, and Warwick & Birmingham Canals in 1800 (these providing an alternative and shorter route between Birmingham and London) Braunston prospered as a canal centre, a fact reflected not only in the volume of boat movements and toll income[4] but also in the decision by Pickford's, in the 1840s, to transfer thirty families from Manchester to a boat repair dock alongside its Braunston depot.[5] Although Pickford's closed this dock, and withdrew from nearly all its canal operations in 1847, the census of 1851 still identified sixty of the 264 inhabited dwellings at Braunston as housing canal workers and 173 out of the population of 1,253 as 'boat people who passed during the night'.[6] In 1858 the area around the bottom lock (on the Grand Junction Canal) was described as a boatman's village 'where the people on the land seemed to belong to the people on the water; where everybody knew everybody and seemed glad to see everybody, and where there was some provision made for [all] the boatman's requirements'.[7]

Despite the challenge from the railways in the second half of the nineteenth century, Braunston retained its position as a major canal centre. Two factors contributed to this. The first was the physical dimensions of the canal locks immediately to the north and south: on the line of the Grand Junction the locks were wide and could accommodate two narrow boats at a time, thus facilitating boats working in pairs, whilst on all the canals between Braunston and Birmingham the locks were narrow, could accept only one narrow boat at a time, and were generally worked by single boats.[8] As a result Braunston became a key

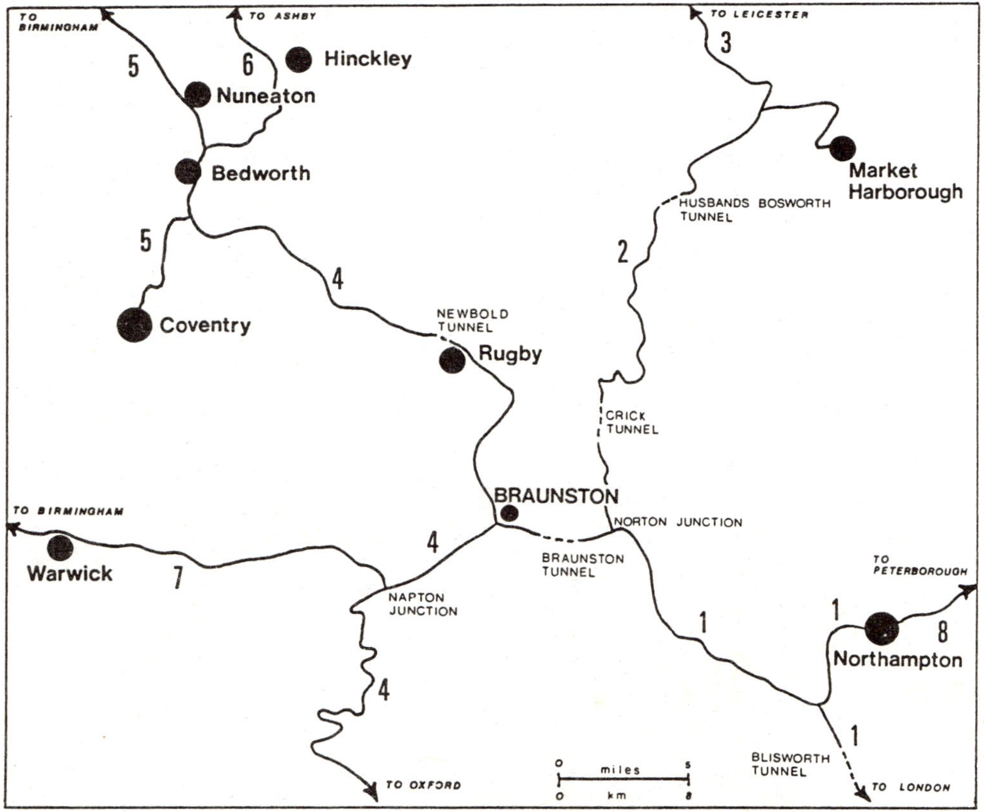

Fig. 1. The position of Braunston. *1* Grand Junction, *2* Grand Union, *3* Leicester & Northamptonshire Union, *4* Oxford, *5* Coventry, *6* Ashby, *7* Warwick & Napton canals, *8* river Nene.

change-over point and attracted extensive wharfage, stabling and warehousing as well as boat-building and repair facilities.[9]

Second, although a succession of different companies dominated long distance trading between London and the Midlands throughout the nineteenth century, there was a tendency for each company to absorb the staff, vessels and established customs and practices of its predecessor; the effect was to maintain the role of those existing depots on the waterway network.[10] By the 1890s the most important canal carrying company trading on this route was Fellows Morton & Clayton, a firm whose origins were in the industrial west Midlands.[11] Its take-over of the London traffic of the London & Midland Counties Canal Company in 1886, by other mergers and a programme of boat building which saw over 140 vessels added to the fleet between 1886 and 1923,[12] gave FMC sufficient importance on the Grand Junction Canal for it to negotiate preferential tolls with the Grand Junction Canal Company and to act as its intermediary in the purchase of the canal branch to Leicester in 1894.[13] FMC had several depots on the waterway system, but on the main route of the Grand Junction Canal, where its growing presence was to help increase traffic tonnages by over 50 per cent between 1888 and 1905,[14] it had only two — at Brentford and at Braunston. Thus in a period of general decline in canal carrying the emergence of FMC as one of the largest canal-based companies in England was to retain the status of the Grand Junction Canal as a major trade artery and Braunston as one of its most important centres:[15] indeed, to generations of boatmen employed by FMC 'the village of Braunston was regarded as home [and] . . . they wanted to be buried here whether they had a home here or not and wherever they died'.[16]

The origins of the strike

Disputes between canal boatmen and their employers were not unknown during the nineteenth century and were restricted neither to certain parts of Britain nor to particular companies. Nevertheless, there is little to suggest that such disputes, including strikes, were precipitated and organised other than by the boatmen themselves and then only by those working for the larger carrying companies.[17] Evidence presented to the Factory Commission of 1876 indicated that 'canal boat drivers have no organisation amongst themselves like a Trade Union',[18] though this was equally true of many other unskilled and general workers employed in the transport industry. In this sense the canal boatmen should not be seen as a special case, though the peculiar circumstances of long-distance boating rendered their occupation particularly difficult to organise into concerted industrial action.

The unionisation of the boatmen working for FMC is unclear, since detailed union and company records appear not to have survived.[19] What is clear, however, is that much of FMC's trading activity was concerned with the movement of cargoes between England's major ports and the industrial Midlands: inevitably the loading and unloading process at the docks brought boatmen into contact with lightermen, rivermen and other dock workers. Particularly significant in this respect was the contact at Brentford and Limehouse with the London dockers,[20] for it was among this group that men such as Ben Tillett and Harry Gosling had demonstrated the strength of organised labour — *vide* the London dock strike of 1889 — as a device for tackling working conditions and other occupational grievances. From such successes in the docks these union leaders devised a twofold strategy, first to amalgamate the separate, but transport-related, unions into one powerful body 'as a means of giving effect to the claims formulated by the port workers, by the seamen and other bodies of workers in the transport trade',[21] and second to draw public attention to the iniquities of their working environment by mass meetings and strikes and by pressure on government to establish investigative commissions and inquiries.

In 1910 over thirty unions engaged in waterside transport work agreed to create the National Transport Workers' Federation, with Harry Gosling as President,[22] and there followed, in 1911 and 1912, a succession of strikes which, on occasions, held up all services borne by rail, road, canal or river throughout the country. Although there were difficulties in achieving co-ordinated industrial action, government controls over rail and canal transport during World War I brought the unions closer together and quickened the process away from federation and towards amalgamation.[23] In June 1920 the National Transport Workers' Federation voted in favour of an amalgamation of transport unions, and there followed discussions and conferences to which unions representing road transport workers were invited. Following balloting of individual members, eleven unions met in May 1921 and decided to merge into the National Union of Transport and General Workers, which would officially function from 1 January 1922: Harry Gosling was elected president, and secretary to the Waterways Trade Group, Ernest Bevin secretary both to the union and to its Docks Trade Group.[24]

Associated with this process of political reorganisation was a growing commitment among the union leaders to the plight of all other general workers in the transport industry. In particular the canal boatmen were identified as one minority group requiring special help, for the union 'keenly realises that there are many things from which these people have to suffer and some of them can only be described as scandalous'.[25] Early in 1919 a joint committee of employers, and representatives of the Transport Workers' Federation, was formed to consider the hours of work of canal workers, whilst in November 1919 Ernest Bevin, on half of the Federation, achieved 'the appointment of a committee, composed of public men together with departmental representatives to inquire into the whole subject [of living-in] on canal boats'.[26] The living conditions of canal boatmen were revealed to a wider audience early in 1923 when, as the newly elected MP for Whitechapel, Harry Gosling posed questions to the Minister of Health,[27] and also later in the same year when he became Minister of Transport and Paymaster General in the first Labour government.

This thrust at national level was complemented by vigorous local recruitment campaigns, and by union intervention in localised disputes about wage rates and working conditions. Particular efforts to increase membership were made at Kidderminster and Bilston in the Midlands, Ellesmere Port in north-west England and, in association with the union's Docks Trade Group, at Brentford, 'where this important junction between the Midland canal system and the Port of London is becoming a strong outpost'.[28] It will be recalled that FMC had an important depot at Brentford, so it must have been here that boatmen working for the company's southern fleet were recruited. Between September 1922 and March 1923 the union had successfully negotiated agreements on wages and conditions of employment with several canal companies, such as W. Gossage (Widnes), the Midlands & Coast Canal Carrying Company, the Severn & Worcester Canal Company and the Birmingham Canal Navigation Company.[29] However, in March 1923 there was the first indication of problems when the TGWU journal *The Record* stated that 'difficulties have been experienced in conducting negotiations with certain firms engaged in canal transport, particularly [those] on long journeys'.[30]

The root of the problem was the union's determination to convert its piecemeal negotiations with individual companies into a national programme for canal workers. By June 1923 this programme had been approved by the TGWU's General Executive Council. It contained six major proposals:

1 The working week would be based upon forty-eight hours, and where a shorter working week was in operation this would continue but minimum wage rates would be agreed.
2 A joint committee would be appointed to calculate the time for each working trip and to fix tonnage rates.
3 There would be a weekly wage for captains of £1 5*s* and of £1 for mates, together with tonnage rates as agreed in (2) above. For steam-powered narrow boats, captains would be paid £1 2*s* 6*d* and drivers £1, with a bonus for the latter of 2*s* 6*d* for washing boilers. Tonnage rates would be added to these basic wages.
4 The agreed rates would not be subject to any deductions for provisions for horses, or for provender, gear and other extraneous charges.
5 Within twelve months of the date of this programme, two adult males would be employed on each boat, both of whom would be recognised as company employees and paid as such.

6 A sub-committee would be formed to inquire into the wage rates of other canal employees such as butty boatmen, lock keepers and tradesmen. This would make recommendations on a minimum wage level.[31]

Underpinning this programme was a desire to remove two fundamental problems from the established system of long-distance canal carrying. The first, a social problem, concerned the employment of women and children on narrow boats. Despite earlier attempts to regulate living conditions on boats and to provide an education for boatchildren,[32] the 1921 inquiry, prompted, it will be recalled by the TGWU's predecessor, had still revealed a substantial number of family boats on the canal system and high levels of illiteracy among both parents and children. Thus, argued the union, if crews became all-male, wives could be left at home and children provided with a continuous education. Furthermore such action would remove an anomaly in the Workmen's Compensation Act of 1897 whereby women and children were not covered for sickness and injury, since it was argued that only the boatman worked for the employer and that his dependants worked for him. However, since wives and children provided an important source of unpaid labour, their removal would need to be accompanied either by an increased payment to the boatman or by additional paid labour. A reform of the wage system, therefore, became the second major thrust to the union's programme.

It is virtually impossible to identify any real pattern to the wage structure for working boatmen, since each company had its own policy, with complex variations based upon route, amount of night travelling, tonnages carried, war bonuses, cargoes, duration of trip and tasks, and expenses incurred. However, from contemporary newspaper reports of interviews with FMC boatmen it is possible to get some indication of income levels. One account, of a pair of boats working from Birmingham to London with a crew of three men, recorded a gross income of £6 7*s* 6*d* for the fifty-five hour journey, out of which there were expenses of about £1 for the purchase of ropes and tunnel tickets, whilst in another a boatman carrying 25 tons from Coventry to Brentford, and returning to Wolverhampton with a 'back-load' of 24 tons, received £9 10*s* for a trip which lasted twenty-three days. When the latter had paid all his expenses he was left with about £6 or under £2 per week to keep himself, wife and three children. FMC estimated a weekly income of £10 15*s* for a man, his wife and two youths working a motor and butty boat between south Staffordshire and London, and £5 3*s* 9*d* for a single-horse boat on the same journey. However, because boatmen were not paid for waiting time (or were given only a demurrage rate of 2*s* 10*d* per day) and were not guaranteed a back-load, average rates were most unlikely to reach these figures. Indeed, one boatman stated that 'taking the year through, we work twelve hours a day, seven days a week and we do not average 25/- per week'.[33]

On the basis of these grievances the TGWU called a national conference with the major carrying companies, but the companies were reluctant to participate. First, as a director of FMC had stated two years previously, the introduction of the forty-eight-hour week would mean 'it would take so long to get your boats through the country that you will practically have nothing to carry . . . we are in competition with railways . . . [and roads] . . . and time is an element of the [price] contract [with the boatmen]'.[34] Indeed, the operation of such a scheme with the Shropshire Union Railway & Canal Company was cited as the main reason for the withdrawal of this company (with over 200 boats) from all canal carrying in 1921.[35] Second, the removal of unpaid wives and children would lead inevitably to an increased wage bill for the companies and a further loss of competitiveness with road and rail transport. For the latter part of World War I all independent canal companies had been placed under the Canal Control Committee, which, among other things, had restored any financial losses incurred. The decontrol of the independently owned canals in 1920 meant a return to open competition and, with the economic recession of the early 1920s, this had led to cuts in staffing levels and in wage rates.[36] Unlike most canal carrying companies, FMC was better able to compete with road and rail transport, since it tended to specialise in high-class traffic, such as foodstuffs and metals, which attracted better rates, and could offer a wide range of facilities (warehousing, cartage and clerical staff) more akin to those provided by the railway and road haulage companies. By 1920, however, FMC was facing a 150 per cent increase in its toll payments (compared with pre-war levels) to the canal owners as they in turn struggled to compete with the railways for a diminishing haulage trade; since this meant the FMC had to adjust freight rates to its customers, 'until we have had a twelve-months period working with those, we hardly know whether we are solvent'.[37] Any conference, therefore, which might increase either individual wage levels or the overall wage bill was unlikely to find support among the carrying companies. And because the union proposed the removal of women from boats there was little enthusiasm, for this aspect at least, from the boatmen either. Indeed it was stated that 'they would separate from their Union rather than separate from their wives'.[38]

The refusal of FMC to attend the conference was compounded by its determination to act independently on the wage levels of its employees. Thus when the railway companies reduced rates early in 1923 FMC responded with its own cut (of 15 per cent) in April and then, on 10 August 1923, it announced an intention to reduce boatmen's wages from the following Monday — for steamer captains working between Braunston and Brentford a cut of 4*s* on a round trip, for steamer and butty boat operating as a pair of cut of 10*s*, and for London-bound butty boats and horse boats a reduction of 4*d* per ton on tonnage rates. The average reduction amounted to about 6·5 per cent. This was the last straw as far as the union was concerned, and as soon as Mr Shaw, a full-time official, returned from the FMC offices confirming the company's decision, the TGWU 'had no alternative but to stop the boats'.[39]

The strike

There were about 600 men on the FMC payroll at this time,[40] and, although not all its employees were members of the TGWU, the recruitment campaigns at Brentford and Ellesmere Port seem to have achieved a 100 per cent membership of FMC boatmen. Thus, when the strike call was made, it was natural that the boats would be assembled at FMC depots so that maximum pressure could be placed upon the company.[41] At Braunston between fifty and sixty boats tied up in the approaches to the FMC wharf and along both sides of the Oxford and Grand Junction canals. This created an addition of about 300 people to the village's population (1,081 in 1921) and considerable pressure upon the local facilities. In order to administer the strike at Braunston the TGWU sent Mr Sam Brooks from Area 5 headquarters at West Bromwich. He took up residence at the Ship hotel, adjacent to the wharf and to FMC's depot. Despite minimal experience of the canal trade,[42] Brooks was to make a big impression on the boatmen, not just by his physical appearance — for he was a tall, well built man

Fig. 2. The FMC depot at Braunston, a photograph taken during the boatmen's strike from the cast iron bridge seen in fig. 3 (William Knibbs, by courtesy of Alan H. Faulkner, Colchester).

Fig. 3. Narrow boats *Denmark* and *Australia* at Braunston during the strike. Standing on the fore end of the *Denmark* and wearing the trilby hat is Sam Brooks of the TGWU, who was based at Braunston for the duration of the strike (Ike Argent, by courtesy of Alan H. Faulkner, Colchester).

— but also for the way in which he organised a wide range of social activities in addition to his more usual union duties. Throughout the strike there were concert parties and open-air church services for the strike-bound families, and most functions concluded with the distribution of cigarettes to the men, sweets to the children and a speech from Mr Brooks reminding them all of the need for a determined and united stand until a settlement was reached. There were further opportunities for solidarity, and communal grief, at the three funerals which occurred during the strike, none more so than that for Edward Walker, aged twelve who fell from his father's butty boat and drowned: 'an extremely impressive sight was presented as the cortege, numbering probably 100, proceeded from the Castle Inn, where the body had been resting, to the church . . . many of the followers carried touching bouquets of wild flowers to place on the coffin'.[43]

The continued presence of such a large number of families enabled the union to activate elements of its national programme, particularly those relating to the educational needs of parents and children. Dismayed by men who could not sign for strike pay, Mr Brooks undertook to teach them to read and write, and when Braunston school reopened for the autumn term forty-six boat children aged four to eleven, 'temporarily living at Braunston as their boats are held up by the Boatmen's Strike',[44] were admitted. Although the school had a good reputation for its attention to boat children,[45] it was incapable of absorbing so many into its existing structure, and so, on 17 September, a supplementary teacher, Miss Mabel Hayes, was appointed with a specific responsibility for these children. The school also attended to other aspects of their welfare, and there were several collections of footwear and clothing 'for these poverty ridden children: their boots and underclothing were in a terrible state [with] many young children being without a shirt'.[46]

However, the prime function of the union was to support and represent the boatmen. At Braunston the men were paid full strike pay, regardless of their period of membership, whilst donations and collections produced a further £110 for distribution. This gave a weekly supplement of 2*s* 6*d* to married and 1*s* 6*d* to single men.[47] As far as negotiations were concerned, much activity in the initial stages took place at a national level. Thus, on 16 August, a meeting was convened between Messrs Gosling, Crump and Shaw, representing the Union, and FMC, at which the union offered a resumption of work in return for independent arbitration on the boatmen's case.[48] This FMC refused and, at a later meeting, threatened liquidation of the company if the wage reductions were not accepted. As a compromise, however, it did suggest that the reduction could be carried out in two stages, that there would be no more wage reductions before March 1924 and that it would consider arbitration (and make any outcome retrospective) if the boatmen would resume work on the reduced rates. Meetings at Braunston and at the other depots rejected both proposals. On 21

August there was another meeting between Harry Gosling, A. J. Ash (managing director of FMC) and Sir David Shackleton (chief adviser at the Ministry of Labour) but Ash 'declined to discuss the dispute in a reasonable way'[49] so no further official contacts were proposed for the immediate future. On 13 September all TGWU officers involved in the strike met, and on the following day Harry Gosling travelled to Braunston to speak to the boatmen. His address was of a general nature, emphasising the need for the organisation of canal workers in order to improve working conditions.[50] Specifically he emphasised the importance of regular schooling for boat children, and he made a special point of talking to them at Braunston. In his autobiography he recalled the occasion: 'when I got among the children and they heard who I was, they began to enquire if the strike was over. I told them I was very sorry, but it was not over yet. They did not appear to mind in the least, though I thought it would be bad news for them. In fact they seemed relieved, so I asked why they wanted to know. They told me enthusiastically that because of the strike they had been able to go to school!'[51]

When a further meeting between the union representatives and A. J. Ash did take place it brought no changes whatsoever in the company's position. As a result the union checked whether FMC had made any official move towards its threat of liquidation. This proved negative, though additional evidence of the company's hardening attitude emerged when, in October 1923, all the striking boatmen were sent a letter from FMC's solicitors containing 'formal notice that as you ceased work without proper notice ten weeks ago, they [the company] require you to forthwith quit the cabin which you occupied as their servant'.[52] The union instructed the boatmen to stand firm.

It would appear that the main reason behind the eviction threat was FMC's desire to unload the cargoes on to lorries and to deliver them to anxious customers by road. At Braunston alone it was estimated that 1,000 tons of sugar and tea, bound for Birmingham, were held up. Inevitably the boatmen saw this as an act of extreme provocation and at several of the depots they took steps to impede the unloading operation. At Braunston, the three boats which FMC had selected were moved from the wharf to the canal proper and the entrance was blocked by four other boats loaded with pitch. FMC's officials attempted to gain access to the boats, but a tug-of-war developed with the boatmen, the latter supported by a crowd of womenfolk who 'cheered them and uttered such witticisms as they could think of and deemed appropriate to the occasion.' In the stalemate which followed, a local truce was called in the hope of a national settlement, but when this proved fruitless a second attempt was made to unload the three boats. On this occasion, 5 September, a detachment of police was drafted in to supervise the operation, and after some spirited gestures of resistance by the boatmen, celebrated by the hoisting of flags, the playing of gramophones and melodeons, and dancing and singing on the towpath, the boats were brought to the wharf and successfully unloaded, 'whilst the canal workers sat on the barges [*sic*] near by and contented themselves with booing and shouting'.[53]

The company's determination to distribute strike-bound cargoes to its customers led to similar confrontations at other depots, though in the process it raised a fundamental issue of trespass which was to take the dispute, initially, to the High Court, and to an eventual settlement by arbitration.

During the early years of trading on the Grand Junction Canal arrangements for the safe delivery of cargoes were extremely casual, with the result that the carrying companies either had to reimburse customers for losses in transit, or had to make disclaimers on their responsibilities for any items of value carried on their boats.[54] Persistent pilfering by Pickfords' boatmen, 'a vile set of rogues', compelled that company to make the boatmen totally responsible for a cargo from the point of loading to the point of discharge, a decision upheld in the Court of Exchequer in 1844.[55] Since the succession of companies which dominated trading on the Grand Junction Canal in the late nineteenth century had tended to retain the practices of predecessors, this arrangement still held in 1923. Thus when FMC officials attempted to board and unload the boats they were accused of breaching this long established principle (for, strictly speaking, the boats were still in transit) and of damaging towing and mooring ropes. It will be recalled that the boatman was responsible for the purchase of ropes, etc., so they belonged to him and not to the company.[56]

The union was quick to put forward these points in its defence of the boatmen, though when one of its own full-time officials, Fred Potter,[57] was seen on board FMC boats at Brentford the company was equally quick to repond and, on 6 November, issue a writ against him. This took the form of an injunction

> restraining the Defendant [Potter] from procuring or attempting to procure a breach of contract between the Plaintiff [FMC] and their Bailors by threats of violence or otherwise and from interfering or attempting to interfere by threats or otherwise with the rights of the Plaintiffs to enter into or continue such contracts or contractual relations with the Bailors as they will. And generally to carry on their business as they would and to restrain the Defendant from trespassing or ordering or inducing others to trespass upon the said Canal boats the property of the Plaintiffs.[58]

When the case was brought before Mr Justice Romer on 9 November 1923 there was an immediate request for an adjournment from Mr T. R. Hughes, KC (appearing for Potter) since he had been given little time to consider the evidence. This request developed into a general discussion on the background to the case wherein all the familiar grievances were aired. The significant development, however, was that the union's desire to settle the dispute was now reciprocated by FMC, as a result of which the judge suggested that the case should rest for one week 'in the hope that I shall hear nothing more of it, and that the parties will be able to arrange their differences in the meantime'.[59]

In the event the court did reconvene, though in the intervening week there were active discussions between FMC and the TGWU resulting in a union offer of a return to work, on the terms which applied when the strike began, if a conference could be arranged to find a settlement or if FMC would agree to arbitration. FMC's response was to accept arbitration, and so the court's function was to define terms of reference, namely that the Minister of Labour would appoint an arbitrator 'who shall, after hearing the parties, decide the following question: whether or not the employers are justified for the reason mentioned in making the proposed or any reduction in the wages of the said employees'.[60] This outcome was relayed to all the depots by the local representatives, and at the Braunston meeting on the evening of Sunday 18 November it was agreed that the union's recommendation of a return to work be supported. At about 7 a.m. on 19 November 1923 'the boats began to move, but some of them may be delayed several days longer owing to the fact that horses require attention before resuming their journeys on the towpath'. On Friday 23 November Miss Hayes ceased her duties at Braunston school and by the following Friday all the boat children had been withdrawn from the school and had rejoined their parents' boats.[61] The strike at Braunston was over.

Conclusion

Arbitration on the case took place at the industrial court which met in London on 20 November. After hearing evidence from both sides it decided that 'the average reduction of 6·47 per cent proposed by the company went further than was necessary', though it expressed the opinion that, with the revival of trade, FMC should be capable of paying a fair remuneration to its employees, particularly those working the lower-paid horse-boat traffic. Hence the court directed 'that the scales of pay in force immediately after the last adjustment in January 1923 be readjusted by an average reduction of 5 per cent, to take effect in two equal instalments of 2·5 per cent each, the first reduction to take effect on single trips commencing on and after 19 November 1923 and the second reduction on single trips commencing on or after 19 December 1923'.[62] Other tonnage and trip rates were to be adjusted on a similar basis, with any item upon which the parties could not agree to be referred back to the court for final determination.

Despite the fact that the boatmen had to accept a wage cut, the union saw the dispute, including the outcome, as a major success. First, it had sustained the strike for three months in a number of geographically dispersed depots without a single boat being moved. This was a tribute not only to the boatmen (and women) but also to the union's organisational ability and its regional and national officials. Second, the fact that the union had denied FMC's intention both to reduce wages without consultation and to refuse arbitration was a success for its strategy in handling relations with employers. Under the guidance of leaders such as Ernest Bevin and Ben Tillett, all unions had to accept that their role in the difficult economic circumstances of the early 1920s was little more than that of 'protective societies concerned with resisting wage reductions'. Nevertheless, argued Tillett, such conditions 'tend to impose upon both parties [i.e. unions and employers] a commonsense policy of co-operation, admitting duality of authority but insisting upon a unity of interest'.[63] In such circumstances negotiations and direct formal relations with employers became the key elements of the TGWU's strategy — hence the national programme for canal workers — though, in a post-war period which witnessed rising unemployment, a fall in real wages, increasing frustration at unfulfilled social problems and a negative attitude on the part of employers, it was realised that the strike weapon still had a powerful cutting edge. Thus the policy for the first few years of the TGWU's existence was to employ peaceful negotiation and strike action as two different yet complementary approaches 'provided priority was given to negotiations'.[64] The course and outcome of the boatmen's dispute was, therefore, a clear vindication of this strategy. Third, the successful organisation of this dispute gave the TGWU much-needed confidence after its disastrous dock strike earlier in 1923. In this dispute Ernest Bevin had successfully negotiated a two-stage wage reduction for dock workers, only to see a strike of 40,000 against the agreement just before the second stage was to become operational, with one group of workers, led by Fred Potter, seceding from the union.

Bolstered by the success of the boatmen's strike, the TGWU turned to other aspects of its national programme for canal workers, with Harry Gosling, president of the union and Member of Parliament, as its leading campaigner. The anomalous position of the boatman's wife, under the Workmen's Compensation Act 1897, was removed in 1924 when FMC and Thomas Clayton (Oldbury) Ltd agreed to be responsible for the employment of a captain and mate on each boat, but attempts to legislate for the findings of the 1921 'Living-in' committee — particularly those relating to the employment of women and the education of children — were not so successful. A private member's Bill, sponsored by Gosling, received considerable attention in the Commons and in the letter page of *The Times*, but by the time it had reached the committee stage in 1931, a year after Gosling's death, there had been so many modifications to its original intentions that it was withdrawn.[65]

The successful penetration by the TGWU into the canal community sent shivers down the spines of the canal proprietors. This was particularly the case on the Grand Junction Canal, where the dock strike and boatmen's strike of 1923 had resulted in considerable loss of income from tolls and tunnel charges.[66] Thomas Millner, engineer responsible for the section of canal from Braunston to Fenny Stratford, had been sent to Braunston on several occasions to ensure the free passage of non-FMC boats, and in mid November he received a letter from William Yates, the Grand Junction Canal Company's chief engineer, requiring information on the 'absolute minimum number of men you can manage with, assuming all the work except what is absolutely necessary is stopped'. Although the union had few members among GJCC employees, Millner was told, 'please go into this carefully for fear of a total withdrawal of labour'.[67]

How much the boatmen gained from the strike is more difficult to assess. The contemporary feeling, expressed by FMC and by a reporter for the *Manchester Guardian*, was that the union alone was responsible for the strike action and that without the behaviour of TGWU officials such as Fred Potter 'there would be no difficulty in coming to terms with the men'.[68] Furthermore, although they stood behind the union to a man, there must have been bitter disappointment at what they ironically described as another 'Irishman's rise', albeit 1·5 per cent less than that wanted by FMC. In view of the failure to achieve very much else of the union's national programme life after the strike continued much as before, but on reduced wages, though to this day there is an enduring feeling at Braunston that some of the strike ringleaders were victimised by the company, either by the switching of some of its operations to other depots or by the severity of Mr (Mac) Anderson, its fleet manager, on men who did not meet his demanding schedules. Nevertheless the company did continue to trade, and through the difficult years of the 1920s and 1930s it provided regular work for its employees — indeed, immediately after the strike it embarked upon a major modernisation programme, phasing out its uneconomic steamer fleet and introducing new diesel-powered boats and unpowered butties.[69] In 1929 the Grand Junction Canal Company merged with the Regents Canal, Warwick & Napton and Warwick & Birmingham companies to form the Grand Union Canal Company; there followed a programme of massive investment in the line of canal between London and Birmingham, including dredging, the widening of locks north of Braunston, and the creation of a new carrying fleet.

However, although there was some revival of the canal trade in the 1930s and during World War II, most of the canals were nationalised on 1 January 1948 and their trading activities taken over by the British Transport Commission. Still a family-controlled firm, FMC found it increasingly difficult to operate alongside this new set-up and against road transport — the latter a particularly suitable form of conveyance for the higher-value products, such as sugar, tea and sauce, in which FMC had specialised. In the first half of 1948 the company recorded its first-ever trading loss, of £5,000, and decided to go into voluntary liquidation. On 1 January 1949 its assets were bought up by BTC and the boats in its southern fleet incorporated with those of the former Grand Union Canal Carrying Company.[70]

With the demise of FMC and the progressive decline in commercial narrow boating on the Midland canal system, the strike of 1923 has become an increasingly distant memory, both for the boatmen who participated and for the union which organised and controlled it. Nevertheless, the strike remained the only one in FMC's history of sixty years of canal carrying, and one of the first disputes to be handled by the newly created Transport & General Workers' Union. It is for these reasons that it will endure and that its story deserves to be told.

Notes

1 Two other companies also became involved in the strike, namely the Midland & Coast Canal Company and the Chester & Liverpool Lighterage Company Ltd. At the end of the dispute these two companies agreed to accept the settlement imposed upon FMC.

2 In the TGWU house journal *The Record*, XXVIII (1958), the obituary notice for Sam Brooks, union organiser at Braunston, gave the year as 1921 and the duration as sixteen weeks. In Harry Gosling's autobiography *Up and Down Stream* (1927), p. 195, he also gave 1921, even though he was actively involved in the dispute, as president of the TGWU and secretary to its Waterways Trade Group, and had spoken to the boatmen at Brauston.

3 G. L. Turnbull, 'Pickford's and the canal carrying trade, 1780–1850', *Transport History*, VI (1973), pp. 5–29. Braunston stood on the London–Chester road, and the stretch between Old Stratford and Dunchurch (near Rugby), which passed through the village, had been turnpiked as early as 1720. There were further improvements to the road by Thomas Telford in the 1820s and this, together with its location at the junction of the Oxford and Grand Junction canals, made Braunston 'a village of great thoroughfare both by land and by water' (Pigot & Co.'s *National Commercial Directory*, 1830, reprinted by Northamptonshire Libraries, 1975).

4 Between 1793 and 1806 the takings at Braunston toll office increased from £966 to £19,040 and the tonnages gauged from 953 tons to 18,495 tons, whilst in 1830 the toll takings at Braunston (£35,365 6*s* 11*d*) were twice those at neighbouring Hawkesbury (£15,482 18*s* 9*d*) and Napton (£13,419 9*s* 2½*d*). H. J. Compton, *The Oxford Canal* (Newton Abbot, 1976), pp. 79–80; Northamptonshire Record Office (N.R.O.), RB 19 Canals.

5 Parliamentary Papers, 1852, XL, p. 298. The move took place between 1841 and 1845 and was completed only two years before Pickford's withdrew from canal trading.

6 N.R.O., Braunston Parish Records, ACC 1974/227.

7 J. Hollinshead, *On the Canal: a Narrative of a Voyage from London to Birmingham in 1858*, Waterways Museum, 1973, p. 44.

8 A. H. Faulkner, *FMC: a Short History of Fellows, Morton and Clayton Limited* (1975). The locks on the main line of the Grand Union, formerly the Warwick & Napton and Warwick & Birmingham canals, were widened in the 1930s.

9 Kelly's directories for the later nineteenth century give the following information on boatbuilding at Braunston: 1854 James Hughes, boatbuilder; 1877 Thos. Bradshaw and James boatbuilders, Wm. Stephenson, boatbuilder, Wm. Nurser, boatbuilder. In 1885 Nurser was the only boatbuilder recorded. This family firm continued until 1941, when it was taken over by the Samuel Barlow Coal Company.

10 Pickford's was based at Braunston from 1788 to 1847 and was followed by the Grand Junction Company's Carrying Establishment from 1847 to 1876: this company employed William Camwell of Braunston for craft repairs and the building of new boats. After 1876 the Grand Junction's traffic manager joined the London & Staffordshire Carrying Company, which also bought several GJ boats and steamers and took over much of the London trade, whilst several GJ agents formed the Midland Counties Carrying Company. Both these firms had been absorbed into FMC by 1886. A. H. Faulkner, *The Grand Junction Canal* (Newton Abbot, 1972).

11 Faulkner (1975), *op. cit.*, p. 4.

12 *Ibid.*, pp. 46–8.

13 Faulkner (1972), *op. cit.*, p. 190.

14 Tonnage figures were as follows: 1888, 1,172,463 tons; 1905, 1,794,233 tons (Faulkner, 1972, *op. cit.*, p. 203).

15 According to the registers of boat inspections held by Daventry District Council, and including the stretch of canal passing through Braunston, the percentage of boats inspected which were owned by FMC was as follows: 1890–94, 12·61; 1900–04, 21·83; 1910–14, 27·71; 1920–24, 63·13. At the 1891 census twenty boats were enumerated in the Braunston registration district, in 1911 twenty-two boats were enumerated in Braunston civil parish, and in 1921 twenty-two boats were enumerted in Daventry rural district, which included the parish of Braunston.

16 N.R.O., RB 19 Canals, 104/5.

17 H. Hanson, *The Canal Boatmen, 1760–1914* (Manchester, 1975).

18 *Factory Commission* (1876), *Parl. Papers*, 1876, XXX, evidence of E. Pamphilon (q. 10527).

19 For a similar conclusion see S. R. Broadbridge, 'Living conditions on Midland canal boats', *Transport History*, III (1970), pp. 36–51, 41.

20 In Faulkner (1975), *op. cit.*, p. 12, there is a photograph of FMC boatmen outside the Six Bells inn, Brentford, where inevitably they would have come into contact with Thames watermen. The management of the Six Bells used to provide its boatmen customers with a free copy of the annual tide book for the navigation of the river.

21 B. Tillett, *Memories and Reflections* (1931), p. 239.

22 V. L. Allen, *Trade Union Leadership* (1957).

23 In 1914, 1,025 miles of canals in railway ownership came under the Railway Executive Committee, and 2,251 miles of independent canals were organised under the Canal Control Committee in March 1917. These committees aimed to provide some national co-ordination of canal carrying and agreements on wage rates. The independent waterways were returned to their owners in August 1920.

24 Allen, *op. cit.*

25 *The Record*, II (April 1923), p. 14.

26 Ministry of Health, *Report of the Departmental Committee appointed to Inquire into the Practice of Living-in on Canal Boats . . . [Committee on Living-in]*, 1921, p. 3.

27 *The Record*, II (June and July 1923).

28 *Ibid.*, January 1923, p. 15.

29 *Ibid.*, September–December 1922, February–March 1923.

30 *Ibid.*, March 1923.

31 *Ibid.*, June 1923.

32 Canal Boats Acts of 1877 and 1884. Section B of the Education Act, 1884, required annual reports to be submitted to Parliament on the enforcement of the Education Acts upon children in canal boats.

33 *Rugby Advertizer*, 31 August 1923.

34 *Committee on Living-in, 1921*, evidence of C. F. Jones, q. 436–7.

35 *LMS Railway Magazine*, 11 May 1925.

36 For example, in July 1920 W. Yates, Chief Engineer to the Grand Junction Canal Company, wrote to T. W. Millner, District Engineer for the section from Braunston to Fenny Stratford, asking for a statement on 'what reductions — inevitable owing to the decontrol of Canals by the Government — can be made consistent with the carrying out of the minimum amount of necessary work to keep the Canal open'. By 13 August 1920 informal notices had been served on thirty-two men working on this stretch of canal. On 19 December 1921 Millner was told to reduce the wages of all 48 hour men by 1*d* per hour and 56–60 hour men by 0·75*d* per hour; on 30 March 1922 he was ordered to reduce wages by 0·5*d* per hour for 48 hour men and 0·25*d* per hour for 56–60 hour men; on 20 June 1923 the wages of all men in the district were reduced by a further 0·5*d* per hour, 'a similar amount to the reduction on the railways'. The effect was that between 1 January 1922 and 2 July 1923 there had been a reduction in weekly wages of 10*s* for 48 and 60 hour men and of 9*s* 4*d* for 56 hour men.

37 *Committee on Living-in, 1921*, evidence of C. F. Jones, q. 486.

38 *Ibid.*, evidence of Rev. W. Ward, q. 840.

39 *The Record*, III (September 1923).

40 This figure was quoted in evidence given to the industrial court in 1923. In 1920 FMC had 208 boats registered as dwellings, of which twenty-one were steamers, twenty-five motor boats and 162 unpowered butty boats (*Committee on Living-in, 1921*, evidence of C. F. Jones, q. 279, q. 288. This compares with the 202 boats registered as dwellings in the Shropshire Union company's fleet, 1920.

41 The narrative of events at Braunston has been based upon three local newspapers, the *Rugby Advertizer*, the *Northampton Mercury* and the *Northampton Daily Echo*. The other Midland depots where boats gathered were Birmingham, Wolverhampton, Nottingham, Coventry, Leicester and Market Drayton.

42 Samuel Brooks was born in 1880, and after a period of service in the navy he joined Spreckley's brewery at Worcester. In 1919 he became an officer of the National Union of Vehicle Workers, which in 1922 became part of the TGWU. Brooks's main interests were in the conditions of vehicle workers, particularly road transport drivers, and from 1938 to 1944, when he retired, he was Commercial Services Group secretary for Area 5, based at West Bromwich. He died in 1958. *The Record*, XXXVIII (April 1958).

43 *Rugby Advertizer*, 31 August 1923.

44 Braunston School, Admissions Register, 1923. I am grateful to Mr D. Thrower, headmaster of Braunston School, for permission to use and quote from these records.

45 In the evidence of W. G. E. Childs, Inspector of Canal Boats, to the *Committee on Living-in, 1921*, the school at Braunston was described as 'the best type on the main canals. I do not know who is responsible for the erection, or how it is controlled, but it is a very well-known school along the waterways, and that is where the children most assemble for school' (q. 997).

46 Braunston School, Log Book, 17 September and 5 October 1923.

47 *The Record*, III (January 1923).

48 TGWU, Minutes of Waterways Section, 18 October 1923.

49 *Ibid.*, 21 August 1923.

50 *Rugby Advertizer*, 21 September 1923.

51 H. Gosling, *Up and Down Stream* (1927), p. 195.

52 *Rugby Advertizer*, 2 November 1923.

53 *Ibid.*, 7 September 1923.

54 *Northampton Mercury*, 28 August 1802.

55 Turnbull, *op. cit.*, pp. 18–19.

56 Giving evidence to the *Committee on Living-in, 1921*, C. F. Jones stated that although the men had to buy towing lines and ropes, FMC had offered to sell them at pre-war prices (q. 405).

57 Along with his left-wing colleague Fred

Thompson, Potter had led the dockers' rebellion against Bevin's negotiated reduction of dockers' wages earlier in 1923. Though the TGWU eventually brought them back into the union fold, they continued to criticise Bevin's refusal to use more militant tactics and, after the General Strike of 1926, created a new breakaway union, the National Union of Transport & Allied Workers. This union failed to break the resolve of Bevin or of the TGWU, and after October 1926 nothing more was heard of the union or of Potter and Thompson. A. Bullock, *The Life and Times of Ernest Bevin*, I, *Trade Union Leader, 1881–1940* (1960).

58 Public Record Office (P.R.O.) J54/1874. Potter was asked to pay 1*s* by way of damages and £135 as costs to FMC (P.R.O. J15/3693).

59 There is a full verbatim report of the court case in *The Record*, III (November 1923).

60 *Ibid*., December 1923.

61 *Northampton Mercury*, 23 November 1923; Braunston School, Log Book, 23 November 1923.

62 *The Record*, III (December 1923); *Rugby Advertizer*, 30 November 1923.

63 Allen, *op. cit*., p. 48, Tillett, *op. cit*., p. 247.

64 Allen, *op. cit*., p. 94.

65 *The Record*, IV (November 1924), and see also XI (August 1931); H. Hanson, *Canal People* (Newton Abbot, 1978), pp. 136–40.

66 Analysis of the monthly statements of costs incurred by the Braunston tunnel tug reveals the following: boats towed through Braunston tunnel (August to November) in 1922, 2,891; 1923, 2,499; 1924, 2,947; profit (August to November) in 1922, £124 6*s*; 1923, £83 2*s* 5*d*, 1924, £175 10*s* 6*d*. In other words the Grand Junction company lost about £70 profit on the Braunston tug alone during the strike; however, despite the strike, nearly 2,500 boats were towed through the tunnel, an average of twenty-eight per day, indicating both the considerable volume of traffic still using the canal and the intention of the company to keep the route open. (I am grateful to Mr George Freeston, of Blisworth, for the loan of these returns.)

67 Letter from W. Yates to T. W. Millner, 12 November 1923. Information from private papers belonging to Mr George Freeston.

68 Hanson, *op. cit*., p. 158. In the *Manchester Guardian*, 14 September 1923, it was stated, 'This is their first strike, and is due to the accident that they now belong to the Transport Workers Union.'

69 From 1923 to 1929 forty-six motor boats were commissioned for the FMC fleet, and a further fifty-three between 1930 and 1939. Faulkner, (1975), *op. cit*., pp. 46–8.

70 C. Hadfield, *The Canals of the East Midlands* (Newton Abbot, 1970), p. 247.

11 *Dutch inland transport in the nineteenth century: a bibliographical review*

H. J. DE JONG

The development of the infrastructure and of transport in the Netherlands in the nineteenth century has been seen as closely related to the general industrial performance of the country in both past and recent writings on economic history. Most economic historians have laid stress on the lateness of mechanisation in industry, starting in 1860 (and on a large scale only from 1895) and on the lack of economic integration, compared to neighbouring countries like Belgium and Great Britain. Two central questions dominate this debate. First, there is the question of the relation between development of the infrastructure and the industrial backwardness of the country. It has been argued that the high transport costs of inputs (the country lacked indigenous coal and iron ore), a poorly integrated market and the lack of economic opportunities due to an inadequate level of transport services retarded Dutch industrialisation. R. T. Griffiths summarised the point: 'Indeed the general opinion among historians of the Netherlands is that the backward state of the transport system in general, and railways in particular, was a major contributory cause of the country's industrial backwardness.'[1]

Secondly, the question has been posed of the extent to which the Netherlands were already endowed with a favourable system of natural and artificial waterways, canals, rivers and open water before the coming of the railways. If the Netherlands were in fact endowed with an adequate system, this would weaken the central argument of the first question and furthermore might partially explain the late construction of the railways. Although the first line was completed in 1839, only from 1860 onwards did a national network emerge. In 1850 the country had only 176 km of railway, whereas in Great Britain and in Belgium there were 9,797 and 854 km respectively.[2] But then, would the waterway system have been sufficient to allow earlier industrialisation?

The aim of this article is to review the literature dealing with these two questions. I will focus first on the size and quality of the infrastructure and on the various transport modes that were in existence around 1800. Next I discuss the coming of the railways and their relation to the waterways and the roads in terms of traffic and transport: to what extent were they competitive or complementary? Thirdly and lastly the question will be addressed as to why the railways were so late and what the effects of the delay were on the economy.

Barges and passenger transport

Although the Netherlands were well endowed with natural waterways, the question has been posed whether they formed an adequate and integrated system, well suited to the needs of inland transport. Griffiths speaks of a popular myth that the early development of the Netherlands and the late development of the railways both have their origins in this waterway 'network'.[3] At the beginning of the nineteenth century the country had many different kinds of waterways: rivers, canals and inshore waters. However, the general view nowadays is that most waterways were of only local importance, not well integrated, as many of the canals had been built primarily for drainage and the exploitation of peat deposits.[4] Let us take a closer look at the system.

At the start of our period a real transport network already existed in the form of the *trekvaarten*, canals with a towpath, on which horses hauled barges to provide regular scheduled passenger services. Heavy freight traffic on these narrow and shallow canals was forbidden by the municipal authorities and anyway was technically almost impossible. The system had been built by the seventeenth century and had not changed much since. In *Barges and Capitalism* Jan de Vries has extensively analysed the economic impact of this transport system on the Dutch economy.[5] Thirty-nine separate routes consisting of more than 600 km of canals connected all the larger cities in the west and north of the country. De Vries has estimated the total capacity of the passenger transport services by *trekschuiten* (barges) at 80 million passenger-kilometres per year in the second half of the seventeenth century, about 68 per cent of the total passenger transport capacity. The passenger volume carried on this network declined sharply in the eighteenth century and by the advent of the next century (1800–06) it was about 22 million passenger-kilometres.[6] This system performed a very specific function in the economy; it was meant only for passenger transport and the barges had limited space for goods. Merchandise and freight had to be transported by other waterways, for instance the larger canals and rivers or the sea.

Coastal shipping on the Zuider Zee

The *trekvaart* network of the western provinces was linked to the south via the estuaries near Rotterdam, Dordrecht and Zeeland (the so-called Delta). It was linked to the northern network by the large Zuider Zee (now the IJsselmeer). This large bay served as a link between five Dutch provinces, a sort of giant traffic interchange. Unfortunately little quantitative research has been done into navigation on the Zuider Zee, let alone into its economic impact. There is, however, strong evidence that the traffic was very dense between a dozen cities that served as gateways to their provinces. A large proportion of the traffic in the bay can be ascribed to the shipping of peat. Van Schaik shows that from the fifteenth century onwards peat was transported over long distances, from the provinces of Friesland, Groningen and, in the nineteenth century, Drenthe to Holland and Zeeland.[7] Navigation on these waters also took place by *beurtschepen*, packet boats carrying freight and passengers between fixed points on a regular schedule. Timetables and destination were regulated by the local authorities, i.e. by agreements drawn up between the administrations of the towns served by this kind of transport. The development of these local monopolies has been described by Fuchs.[8] The precise founding dates of the services are not known, but they were established by the mid-seventeenth century, and the pattern prevailed well into the nineteenth. De Vries studied the timetables and transport taxes of the services. During the eighteenth century, and probably also in the first half of the nineteenth century, there were at least fifty sailings in each direction between the Zuider Zee ports of the provinces of Holland and Friesland (in the north) in a non-winter week. The Zuider Zee also served as a gateway for ocean-going vessels to and from Amsterdam.[9]

The construction of canals and their economic impact

In the 1820s, in the reign of William I, a short burst of canal construction took place in the Low Countries. The economic policy of the government was directed at strengthening the previous trading position of the Northern Netherlands (the present-day Netherlands), especially the staple market of Amsterdam. The Southern Netherlands (now Belgium), which were united with the new kingdom in 1814 (and which seceded again in 1830), were already industrialising and the aim of William I was to unify the two regions by combining the industrial capacity of the south with the commercial character of the north. A very important part of his policy consisted of the restoration of the connections between the Dutch harbours, the sea and the German hinterland.[10] For instance, between 1819 and 1825 the Noord-Hollands canal was built to restore access to international trade for the seaport of Amsterdam. The so-called Keulse Vaart was improved, too, for better transit between Amsterdam, the river Rhine and the German town of Cologne. Access to the ports of Rotterdam and Ghent was improved by means of canals, constructed between 1827 and 1829. Furthermore a new canal was built between the cities of Bois-le-Duc and Maastricht in the south (see fig. 1). Surprisingly this canal was not extended to the rising industrial area of Liège,

Fig. 1. Canal construction in the 1820s. Based on H. Schmal, 'De ontwikkeling van de infrastructuur van het einde van de achttiende tot het midden van de negentiende eeuw', *De Physique Existentie dezes Lands* (Amsterdam, 1987), p. 101

a centre of Belgium's thriving iron industry. All these new canals ran parallel to old natural waterways that were silted up or could not be used any more.

In his study of spatial planning in the Netherlands in the first half of the century Van de Woud states that the building of these canals had clearly been inspired by the English canal works in the eighteenth century, which had proved very profitable. However, there was an important difference in the purpose of the new canals of the two countries. In England the canals gave access to centres of coal mining and iron ore but in Holland construction was aimed at restoring the previous trading position. Therefore Van de Woud speaks of a 'restoration with modern means'.[11] There was a strong belief that the former trading position of Dutch cities like Amsterdam, Rotterdam, Middelburg and Zwolle could be restored by improving their waterway connections. But the new canals did not create economic integration, because they failed to bring about important connections between the various regional waterway systems.

It is not easy to assess the economic impact of this first 'canal boom' because most of the canals did not open up new areas, and therefore did not create new inland traffic. Furthermore there were a lot of shortcomings if the bottlenecks in transport were to be removed. The existing system was characterised by obstacles such as low bridges, small locks, varying depths and narrow dimensions. Sometimes the new canals removed these obstacles but, for instance, the improved Keulse Vaart (part of a connection to Cologne), a 55 km stretch of waterway, still contained seven locks and thirty bridges, all of which imposed a toll.[12] Looking at a map, one might discern a route from Den Helder in the north via Amsterdam to Maastricht in the south. But this route did not become of national importance. There was little call for the movement of goods over long distances, and most vessels were suited only to local or regional conditions. In his exposition of the patterns of traffic and transport in the nineteenth century Kuiler mentions all the different types of vessels that were used in the various provinces.[13] Even a province like Friesland, an area abounding in water, had no direct connection with the waterways of the adjacent provinces of Overijssel and Drenthe, and for that reason interprovincial transport by water was not possible without the help of land carriage or by transhipment.

We must not overemphasise the supposedly minor economic impact of the canals. The building of local canals in the northern area for the exploitation of peat deposits in the second half of the eighteenth century and in the nineteenth, had an important impact on the regional economy. For instance, concentration of people in new settlements, more economic activity and falling transport costs occurred. The shifting of local economic opportunities, however, also gave rise to particularistic and sectional tendencies. There are several examples of outright unwillingness by the local and provincial authorities, supported by private interests, to form connections between these local waterways, for fear of losing existing trading positions. One example is the Drentse Hoofdvaart, a long canal in the northern province of Drenthe, built in 1774. This canal could have served as the linking waterway between the Zuider Zee and both the Ems and the Wadden Sea in the north. But the construction of a relatively short section of no more than 30 km between the town of Assen in Drenthe and the city of Groningen was not completed until 1861. It took so long because the provincial authorities in Drenth feared that trade on such a long-distance inland waterway would harm the provincial economy by by-passing the regional centres. The Groningen authorities feared the canal would threaten the long established central role of the town in the northern economy.[14]

The east of the country lacked artificial waterways altogether. In Twente, an eastern region of the province of Overijssel, which already possessed an extensive rural cotton industry, and which was to become one of the first industrialised regions in the country, there were some small rivers that could be used only in winter. In his study of navigation in this almost waterless area Schutten describes the inventiveness of the bargemen drifting down the brooks.[15] Efforts to improve these waterways failed because of the stubborn attitude of the national government, which judged the proposed canal plans to be of only regional importance and was reluctant to give financial support for construction.[16] So in the supply of raw cotton and the sale of cotton textiles the shippers on these rivers could barely compete with road transport.[17] Although movement by water was cheaper than by road, the latter was faster and more regular. But the cost of transport was not of overriding importance to the domestic cotton industry. In a study of the entrepreneurial Spanjaard family Fischer has estimated the cost of transport of cotton from the port of Zwolle near the Zuider Zee to Enschede near the German border and vice versa at 0·8 to 2·2 per cent of the value of calico in the 1830s.[18]

Nevertheless, the building of the Overijssels canal in the 1850s, between Zwolle and Twente, was of major importance for this region. The canal was part of the 'second building boom' from 1851 till 1860. The second period of building activity was different from the first 'canal boom' and was mainly intended to link the Dutch hinterland areas with the main waterway network.[19] In the case of Twente the new canal not only offered fast and cheap transport of raw materials and finished textile products but also reduced the cost of coal consumed by the steam engines in the spinning and weaving mills that had been mechanised.[20] For example, the price of English coal in Almelo in 1853 stood at ninety guilders per ten tons.[21] The cotton industry areas that were not connected to the canal (Enschede and Oldenzaal) still faced high transport costs, because the coal had to be transported from Almelo by road at about thirty-two guilders per ten tons. For these towns German coal from Ibbenbüren was an alternative. But in 1853 prices were about 174 guilders per ten tons, 117 guilders of which accounted for transport. Small wonder that these towns put their hopes in the construction of a railway to Germany. Boot has observed in his study of the cotton industry in Twente that during the latter 1860s the railway connection with Germany reduced the cost of coal drastically. After 1865 coal prices declined to 100 guilders per ten tons, and this made strong industrial growth possible.[22] Fischer added that speed and reliability were important features of the new transport mode in bringing about rapid mechanisation, which was aided by wage increases, the abolition of a tax on fuel and the falling cost of machines. Fuel costs were about 30 per cent of gross value added in a spinning mill, hence falling coal prices were of major importance.[23] My own view is that in the case of Twente an earlier improvement of the infrastructure would probably have accelerated the process of industrialisation. The same holds true of the southern part of the country: the province of North Brabant lacked waterways, and towns like Tilburg could expand only after the province got its railways after 1860. In fact Tilburg had to wait for a canal until 1923.[24]

Summarising, around 1850 there was no waterway network in the Netherlands but there were several poorly connected local or regional circuits. Despite the active policy of William I and the resulting canal construction, there had been no master plan to furnish the country with a national system of waterways. Apart from new settlements created by canal construction related to peat exploitation, there is hardly any evidence in the literature of new industrial or urban expansion from canal construction before 1850.

After 1865 we discern renewed canal-building activity, according to Griffiths similar to the first boom, and directed at further international integration by improving the accessibility of important seaports: Amsterdam by the North Sea canal; Rotterdam by the Nieuwe Waterweg; Groningen by the Eems canal, which ran to the Ems estuary. Together with the ongoing construction of new inland waterways, like the above-mentioned Overijssels canal, and the connection of local waterway systems this led to the creation of a national waterway network, but certainly not before 1880.[25] The absence of adequate waterway connections in the first half of the century is also the main reason why the introduction of steam shipping was confined to the big rivers and the Zuider Zee. These ships were too large for the existing canals and of course highly dependent on coal, which remained, at least inland, very expensive until the building of railways.

Navigation on the rivers

The big rivers, especially the Rhine, have played an important role in the economic history of the Netherlands. The Rhine was the main trade route not only for colonial products and other staple goods to Germany and Central Europe, for instance coffee, tobacco, sugar, rice, cotton, hides and grain, but also for semi-finished products like pig iron. The most important products that went downstream were grain, stone and especially coal. Important studies on the nineteenth-century development of the Rhine navigation have been written by

Bläsing for the period 1815–51 and by Nusteling for 1831–1914.[26] Especially in this last study the transit trade emerges as a powerful economic activity, furnishing the Dutch economy with a modern service sector. Big structural changes in the Rhine traffic were induced by German industrialisation and by the intense competition between the Rhine navigation and the railways after 1850. The volume of goods transported increased rapidly. The five-yearly average shipped upstream increased from 4 million cwt in 1850 to more than 7 million cwt in 1870. Downstream shipments, destined mainly for the Dutch market, increased from 6 million cwt in 1850 to 28 million cwt in 1870.[27] But important changes had already occurred before 1850. They were due to fierce internal and international competition in the aftermath of the liberalisation of the Rhine navigation established by the Treaty of Mainz in 1831 and also of course to the appearance of steamships. Before 1830 navigation took place by *beurtschepen*, a system comparable to the one we have noted on the Zuider Zee.[28] With the extension of steam shipping the principle of territorial monopoly was broken and by 1845 the *beurtvaart* on the Rhine had almost disappeared. In the rest of the country such monopolies were gradually abolished between 1860 and 1880. According to Kuiler and Griffiths the crucial breakthrough on the Rhine came in 1841 with the construction of iron barges designed to be towed by steamships. This was a cheaper and more efficient means of transport than sailing ships.[29]

Discussing the quality of this route, Nusteling stated that, although the rivers were of major importance to Dutch trade, until 1850 little had been done to improve the navigability of great rivers such as the Lek, Waal, Merwede and Meuse.[30] Shallow water in summer, ice drifts in winter and floods in spring caused significant problems. Efforts to improve the situation were directed not at navigability but at the protection of the land and its inhabitants.[31] Furthermore the control of the rivers was not in the hands of the central or provincial governments but in the hands of the local authorities and administrations, who lacked the money, the knowledge and the equipment to combat the problems caused by the rivers.[32] This situation changed only after 1850, when the government adopted a more centralist approach to the control and maintenance of the waterways (the *waterstaat*) and the principles of a new policy were set out. In a study of the improvement of the tributaries of the Rhine Van de Ven stressed also that the primary concern was drainage. Narrowing the width of the rivers to improve navigability would not remove the danger of flooding. After 1850 the necessity of better drainage as well as of navigability was acknowledged.[33] Competition from the railways and the increased capacity of ships on the Rhine also stimulated later improvements.[34]

Roads in the first half of the nineteenth century

The limitations of the waterways, such as difficult navigability and lack of connections, were obvious to contemporaries. Did this bottleneck lead to an improvement of the transport system in other directions? Let us look at the quality of the roads in the Netherlands and road-building policy in the first half of the nineteenth century. The popular myth of excellent waterways finds its counterpart in the notion that compared to them the roads were of inferior quality. Dutch roads have not received much attention from economic historians. The historical geographer Schmal published an article on the system of high roads in the Netherlands.[35] In *Barges and Capitalism* De Vries dealt with the roads in relation to the system of passenger transport by barges. According to De Vries, at the time of the fall of the Republic in 1795 there were no more than 400 km of surfaced road.[36] More than half of it connected the larger cities with their surrounding villages. About 165 km can be regarded as high roads that connected the larger cities, so they can be seen as complementary to the *trekvaart* (waterway) network.[37] At the end of the French period, in 1814, the total length of these high roads amounted to 450–500 km.[38] Only these main roads were brick-paved, the rest were macadam roads. All other roads in the Low Countries were sandy, not paved, and therefore unsuited to heavy traffic. In summer it was possible for carts with a load of up to 600 kg to use them but at other times of the year it was very difficult to transport bulk goods because of rain, snow and frost.[39] By modern standards this system was of course underdeveloped, but the roads made interprovincial traffic possible by means of postal coach routes. These coach routes either supplemented the *trekschuit* services or served areas that lacked a *trekschuit* network. At the end of the eighteenth century and the beginning of the nineteenth coach services in the region were competing directly with the barges.[40]

After the end of the French occupation, in 1814, some of the high roads were defined as *rijkswegen* (national highways), which meant that the government assumed responsibility for their maintenance. Because of financial problems, however, fewer and fewer roads were favoured with this title, which meant that the financial burden on the provinces increased. The national road system, however, advanced slowly between 1817 and 1850. It connected all the provincial capitals, and this network improved the quality and speed of travel. But one can question the idea that it had any great impact on the mainly regional and rural character of Dutch society, apart from the cities of the provinces of Holland. The amount of investment shows that King William I gave priority to the building of the canals. In 1815 450,000 guilders were reserved for the construction of brick roads on a number of important trunk routes. Although the real expenditure on the main roads was much greater, it never equalled the 20 million guilders that were spent on the construction of canals in the 1820s. Some historians have argued that the real importance of this network lay not so much in the economic as in the political integration of the country. These roads provided for postal communication and were an important symbol of the unity of the young state that had formerly been a conglomeration of autonomous provinces.[41]

Responsibility for the building and maintenance of local roads remained with the provincial and municipal authorities and private individuals. As with the maintenance of the rivers, political and institutional factors retarded road building. Before the constitutional reform of 1848 municipalities had no budgets of their own and so the absence of a road-building policy led to stagnation in local road building. By 1850 not much had changed compared with 1800. Another factor explaining this slow development was suggested by Van der Woud. Road-building materials like bricks and sand could be transported economically only by ship. But most waterways, especially in the east and the south, were inadequate. Van der Woud concluded that the roads were in poor condition not because of the excellence of the waterways but because the waterways were also in a poor state.[42] Comparison of the Dutch situation with developments in other European countries will test the validity of this statement.

Now let us draw some tentative conclusions regarding the economic impact of the roads in this period. In the first place the national trunk roads stimulated coach services. Between 1820 and 1850 the number of daily services between the larger cities expanded rapidly and formed a real competitive threat to the shrinking *trekschuit* services.[43] The growth of coach services was nevertheless short-lived, because the railways were faster, more comfortable and cheaper. Long-distance coach travel disappeared overnight on the routes that were also served by the railway network. However, in the second half of the century these services did not vanish altogether. For instance, the firm of Van Gend & Loos transformed itself into a mail-order house. This company adapted to providing services complementary to those of the railway companies, like parcel delivery, and for business organisations like banks, for instance the dispatch of money, gold and bills.[44] The impact of the roads can be studied in a more quantitative way by using the toll registers of the interprovincial roads. Little research has been done on this topic as yet. Although Schmal has presented some figures on average toll yields between 1830 and 1891 on twenty-one routes, they are not in his opinion very reliable for arriving at estimates of traffic density, because there is no information about how many persons and vehicles were exempt from tolls or could travel on reduced tariffs. His figures show that the intensity of traffic on these roads reached its peak in the years between 1830 and 1850.[45] Construction, however, continued: in 1864 the total extent of surfaced road was 8,542 km and in 1873 12,024 km.[46]

At the regional level there is some specific, but still scattered, information regarding the unlocking of towns from their isolation and their integration into the national economy. For instance Twente, a relatively isolated region in the east of the country, had hardly any surfaced roads before 1827. After this several brick roads, including national highways, were built in the west–east direction, causing a huge increase in road traffic.[47] Although this development dealt a severe blow to the traditional shipping in the area, competition continued between the two modes of transport, because road carriage was very expensive, owing to the tolls, compared with transport by water. In 1858 the charge for transport by road varied between fourteen and twenty cents per ton/kilometre and by ship from one to six cents per ton/kilometre.[48]

After the construction of the first railways in the 1840s and the railway boom of the 1860s and 1870s, some main roads lost their importance. Passenger transport was taken over by the railways and interregional transport of goods went by boat. Toll income on roads declined. The introduction of local tramways from the 1880s was a real threat not only to the main roads but also to local roads. The government economised on the maintenance of the high roads; sometimes the width was lessened, sometimes maintenance was neglected altogether.[49]

Dutch railways between 1830 and 1880

The first general history of the Dutch railways, by Jonckers Nieboer, covered the period 1832–1938.[50] Although the book is more descriptive than analytic it is still of value to railway historians. Not much literature on the Dutch railways has been written in English, the only exceptions being the studies of De Vries and Griffiths.[51] Dutch historiography still lacks a general counterfactual or social savings approach to the railways, as well as any analysis of backward and forward linkages.[52] The absence of reliable quantitative sources has deterred scholars. Furthermore, there has been much scepticism among Dutch historians about the methods of the New Economic History. Of course the economic role of the railways has received attention, for instance in the debate on the timing of the railways, but the quantitative roadbed remains poor. An exception is Van den Broeke's analysis of how the Dutch railways were financed. This book contains an English summary.[53]

The first 'railway movement' in the Northern Netherlands was a reaction to plans in 1831 in the Southern Netherlands (later Belgium) for a railway to connect the port of Antwerp with the rivers Meuse and Rhine, to improve trade with the German hinterland. This so-called 'Iron Rhine' from Antwerp to Cologne was actually completed in 1843. The proposal was seen as an enormous threat to the trading position of Rotterdam, which was developing into a port of transhipment. Amsterdam's position was even more threatened, as the city possessed no adequate link with the river Rhine.[54] The answer was seen as a railway line from Amsterdam to Cologne, and as early as the 1830s the Dutch government acknowledged the importance of the railways' role in improving

Dutch transit trade with Germany. In 1834, however, a subscription to mobilise private Dutch capital for this plan failed, and a proposal thereupon by the government to have the project financed by the state was rejected by parliament in 1838. The arguments against state-financed railway construction were very diverse. Some members of parliament argued against state intervention and exploitation in principle, others boasted of the adequate waterways. Even advocates of the railways questioned the feasibility of raising funds for the project because of budgetary constraints.[55] Hartgerink-Koomans dealt extensively with the public debate on railways at the time, pointing to the rivalry between Amsterdam and Rotterdam, the former city being anxious to secure a direct railway connection with Germany and the latter being the advocate of steam navigation on the Rhine. According to Hartgerink-Koomans their rivalry caused a delay in the construction of the railways.[56] Surprisingly, though, in my view, during these discussions in the 1830s the railways were already being put forward as an important innovation not only for passenger transport but also for the carriage of goods.

Another railway project was well under way in the 1830s: the building and development of a network in the province of Holland by a privately owned company, the Hollandsche IJzeren Spoorweg Maatschappij (HSM, Dutch Iron Railway Company). The main line, between Amsterdam and Haarlem, was completed in 1839. Its route precisely paralleled that of the old *trekvaart*. The line was gradually extended to the larger cities of Leyden, The Hague and Rotterdam (1847). The HSM was almost entirely concerned with passenger traffic and was therefore a serious competitor to the coaches and above all to the *trekschuit*. From 1839 onwards the railway gradually took over the inter-city carriage of passengers from the barges. Van der Knaap has written on the spatial characteristics of network developments in the Netherlands.[57] He suggested that the existing infrastructure in the province of Holland (*trekvaarten* and roads) was reinforced by building routes parallel to the canals for passenger transport and by connecting the urban centres. This, however, so intensified the competition for passengers that railways completely superseded the barges in passenger transport. But the new railway did not compete with the *beurtvaart*, which was involved with the movement of goods and merchandise by waterway.[58]

The operators of the *trekschuit* services tried to fight back by slashing fares. On the Amsterdam–Haarlem route, for instance, fares were reduced from forty to thirty cents. But the operators did not succeed in attracting the same number of passengers as before, because of the enormous time saving offered by the railway. So the great bulk of travellers switched to rail as soon as it was introduced.[59] Passenger numbers rose from 78,000 in 1839 to 470,000 in 1843 and 900,000 in 1849.[60] Given a total population of 3 million the figure is quite spectacular. Because the new line connected existing economic and political centres and replaced an obsolete transport mode, to speak of a railway revolution would be an exaggeration. There was a revolution in speed and therefore in time saving. The journey from Amsterdam to Utrecht took only seventy minutes by train instead of seven hours by barge. Apart from this kind of cost saving, however, there were no fundamental economic changes because this first railway line did not open up new areas.[61]

Despite the parliament's refusal in 1838 to finance an eastbound route to the Rhine, a line from Amsterdam via Utrecht to Arnhem was completed in 1845. It was built with the personal financial backing of the king, who guaranteed a minimum dividend on the bonds issued. In 1845 the enterprise was transferred to private hands as the Nederlandsche Rijnspoorweg Maatschappij (Dutch Rhenish Railway Company). The Rhine railway depended much more than the HSM on goods and merchandise traffic and had to compete with the Antwerp–Cologne railway and with the Rhine navigation. Until the line was extended to the German border its position was not very strong. In 1856 a direct connection with the German railway network was completed.[62] By then Rotterdam also had direct access to this line (1855), so in 1850 the Dutch system was confined to the provinces of North and South Holland, Utrecht and Guelders.

Apart from these two lines, two other small privately owned routes were opened during these years. In the south a railway was built between Maastricht and the German town of Aachen (1853) and another between Rotterdam and Antwerp (1854). In 1855 there were four segments of railway in the Netherlands. They were not connected, were even built to different gauges and were controlled by four different companies. Apart from the two small lines just mentioned, there was no building activity between 1845 and 1860, and the construction of railways in the Netherlands fell behind that of surrounding countries.[63]

Yet during this period there were numerous applications for concessions to build lines in all parts of the country. These plans and proposals are enumerated by Hartgerink-Koomans in two articles on the northern half and the southern half (below the great rivers) of the country respectively. For instance, in the north a west–east railway line had been proposed from the ice-free northern port of Harlingen to the German town of Bremen. In the south a line was proposed from the port of Flushing to Maastricht. These plans for west–east lines saw the Dutch railways primarily as international trade and transit routes and not as a means of integrating the different parts of the country. Sometimes the schemes were inflated to gigantic proportions. For instance, in one of them Harlingen was seen as a central staging post on the British imperial route to Alexandria and India.[64] Some north–south routes were promoted too, for instance to carry imports of English coal from Harlingen in the north and from the south to the rising industrial region of Twente. But there was opposition, particularly from the cities of Rotterdam and Dordrecht, which feared that Antwerp might

Fig. 2. The growth of the railways between 1865 and 1885. Based on H. Knippenberg and B. de Pater, *De eenwording van Nederland* (Nijmegen, 1988), p. 49

gain access to the north–west German market.[65] These ambitious projects failed, however, not from opposition of this kind but for other reasons, which we will discuss. It was not until 1860–85 that a railway system was constructed which was integrated and competitive with the other modes of transport. Lengthy debates took place in several Cabinets, one of which actually fell because of the 'railway question'. Parliament finally decided on a compromise: the construction of the infrastructure would be managed and financed by the state, but the running of the new lines would be left in private hands. The policy was successful; between 1860 and 1885 the Dutch railway system increased in extent from 335 km to 2,392 km.[66] By the end of the period a truly national network connected the emerging industrial towns in the south and east of the country and also the north with the existing lines in the province of Holland (see fig. 2).

The economic impact of the system and its relation to Dutch industrialisation still needs to be examined and quantified. One regional study, however, can be mentioned here. Passchier and Knippenberg have tried to analyse the impact of the railways on industrialisation in the province of North Brabant. Their spatial–statistical analysis shows that industries in towns that were connected to the rail network in the period 1859–74 witnessed a faster growth of total steam power capacity than towns that did not get a railway line. So there is a statistical relationship between railway construction and the development and spread of new mechanised industries in the province. But the authors conclude that transport is only one determining factor alongside others in the pattern of the establishment of industries based on steam power.[67]

In general, passenger traffic was of more importance to the Dutch railways than freight, in contrast to countries like Belgium or Germany. Although the revenue of the railway companies from the movement of goods was increasing, it would never equal receipts from passenger fares.[68] An estimate by Kuiler of the total volume of inland transport shows that the inland waterways were the principal means of moving goods. In 1880 9·9 million tons were carried by the canals and only 4·4 million by the railways, of a total volume of 20·7 million tons. Coastal shipping carried 5·3 million tons. In 1900 the total volume is estimated to have been 56·5 million tons, of which the canals, the railways and coastal shipping respectively transported 28·6 million, 11·7 million and 14·0 million tons.[69]

Before the great extension of the railways from 1860 onwards there had been a competitive struggle between the railways and Rhine navigation over the transport of goods on the west–east through route to Germany. This struggle has been analysed extensively by Nusteling in his book on the Rhine navigation.[70] After 1856, when the railway connection with Germany was completed, a division in the through traffic to and from the Dutch seaports was developing: goods of relatively high value, such as finished products and mixed cargo, were increasingly transported by rail, but raw materials and bulky goods remained with the Rhine ships.[71] For the Rhine navigation the loss of merchandise transport, however, was considerable, so ships became bigger and more efficient, and rates were gradually reduced. In 1868 the river tolls were abolished as a result of the Treaty of Mannheim. Compared to the railways, the Rhine navigation could offer much lower freight rates.[72] From 1868 until the outbreak of the First World War the volume of bulky goods carried by Rhine shipping grew steadily each year. Iron, ores, timber and grain were shipped upstream from Rotterdam to the Ruhr and beyond, German coal was brought downstream to Rotterdam for transhipment. From 1870 onwards the Rhine navigation increased steadily at the expense of the railways. Such a development was accelerated by changes in Prussian railway policy.[73] This flourishing period of the Rhine navigation also changed the attitude of the Dutch government to waterborne transport. The railways were no longer seen as the primary link with the Dutch seaports but rather as a complementary mode of transport. This is clear in the case of Rotterdam. Its growth was due to the through traffic on the Rhine.[74]

Why was railway development so slow?

It is worth considering why railway development was relatively slow as compared with neighbouring countries. One of the most explicit views is that of Jan de Vries, who declared that the late construction of the railways could be attributed to the successful role of the *trekvaart* network in passenger transport. The social savings and the consumers' surplus in relation to a new railway system were very low. The large capital investment required for the introduction of the new technology (railways) was attended by a relatively low cost-reducing impact, because of the existence of an obsolete but still functioning transport network. This would have deterred investors.[75]

This view has been challenged by W. Fritschy, among others, who argues that the *trekvaart* network generated a need for passenger transport and for that very reason promoted the advent of railways. She claims, and I would agree, that at the micro-level investment in a transport system that was to replace an obsolete one was entirely rational, while at the macro-level it may not have been rational at that particular moment, in terms of social savings. According to Fritschy it was not economic but institutional and political factors that retarded the construction of the railways, such as inadequate compulsory purchase legislation and a national policy which frustrated railway construction by a rigidly liberal attitude, that is, no state involvement.[76] The resultant high construction costs had adverse effects on dividends and share prices. Griffiths argues that this deterred investors from putting much capital into these ventures.[77] Jonckers Nieboer has blamed the government for not giving investors dividend guarantees.[78] A quite different judgement was reached by Van de Woud. In his view the late development was fundamentally a problem of planning, as with the canals. The state lacked the organisation to tackle such a complicated matter as the planning and building of a railway network and a canal system, and was therefore unresponsive.[79] There were also specific technical problems such as bridging the major rivers. Lastly, the rivalry, already mentioned, between Amsterdam and Rotterdam, according to Hartgerink-Koomans, delayed the construction of the railways.[80]

All these views explain to a greater or lesser extent the backward state of railway development, often seen as closely connected with the country's late industrialisation. More generally this relationship has been thoroughly discussed in economic history from the 1960s onwards, starting with Rostow, Fogel and Fishlow.[81] On the one hand one might argue that the expected impact of the railway as a transport innovation was too slight (in terms of backward and forward effects and in terms of dividends) to give rise to new economic opportunities. This would be the case if, for instance, existing transport systems were judged at the time to be adequate, or the investment too large in relation to expected earnings to offer an attractive return. The result was retarded economic development and late industrialisation. In short, the delay in railway construction was the cause of the retarded industrialisation. But it is also possible to reverse the causal connection: late industrialisation retarded the construction of railways. So the propensity to construct a national railway system (or any transport system, including waterways and roads) was small on account of the delayed industrialisation.

In the Dutch case this second line of argument seems more plausible. For instance, Fritschy has questioned the idea that the reluctance to invest in railways was due to the inadequate growth-inducing impact of the new system. According to her this notion cannot be a watertight explanation because investors showed great interest in railway investment long before there was any discussion of the growth-inducing impact of railway construction. She refers to the many plans, proposals and initiatives before 1860 that clearly showed interest on the part of industrialists.[82] This group, which was asking for government support, was still too small to overcome the institutional impediments such as inadequate compulsory purchase laws that retarded railway proposals or forced up construction costs. At a national level, before 1855 the demand for railways was not strong enough to ensure earlier construction of a system, although the number of pamphlets and articles published shows that the public recognised the positive influence of railways. The real disincentives to the construction of a national system were gradually removed in the 1850s, when ideas were developed of state-supported railway construction.

This view is more or less supported by the research of Van den Broeke on the financing of the Dutch railways. He demonstrates that although railway entrepreneurs had to display great ingenuity in order to attract the finance required, there was no lack of capital.[83] Most of it came from foreign countries such as Germany, Britain and France, but the supply of capital was never a brake on railway construction by the existing companies. Although the rates of return were low, the expectations were high, and the few Dutch investors who invested in Dutch railways were especially keen on making speculative profits.

The historians Boom and Saal, in an article on railway construction in the first half of the nineteenth century, contest the view that railway development was slow compared with other western European countries. They argue that the pattern of development was a logical result of the economic and geographical character of the country. Because of the late industrialisation, from the third quarter of the century onwards, there was no economic reason to transport large quantities of bulky goods by railway, for instance coal. According to them, the development of the railways was not undertaken until industrialisation in the Netherlands and in Germany showed the necessity and advantages of an integrated transport system.[84] For instance, the railway connecting the ports of Amsterdam and Rotterdam with the German hinterland in 1856 which has already been mentioned came just at the right time. Before then there was no such thing as a European railway network, and therefore no direct competition from large foreign ports. Much the same view is held by Bos, who asserts more generally that the primary cause of Dutch industrial development was the shift in the position of the Netherlands from the periphery to the centre of European industrialisation due to German economic development.[85] This new economic–geographical situation 'forced' the improvement of the infrastructure, that is, of the railways and the waterways, and these developments stimulated the internal integration of the economy and spatially widespread industrialisation. This view affirms, in my opinion, what has been stated earlier on the relation between transport and economic development in the Netherlands: that the development of the Dutch transport system did not bring about any economic revolution but instead went in step with the new economic opportunities that were due mainly to foreign developments.

Conclusion

The nineteenth-century development of the Dutch inland transport system is closely connected with the international transit function of the country. For instance, Amsterdam was still a major staple market in this period, depending on good access to the sea and to the Rhine. At the same time Rotterdam was emerging as a transit port, in close connection with the economic rise of Germany and with the introduction of steam navigation. So government policy concerning the building of infrastructural facilities was largely directed at maintaining and strengthening this trade and transit function. This notion is crucial to explaining and understanding the patterns of inland transport, the contemporary debate on the desired infrastructure, and the transport system that actually emerged in the second half of the century.

Dutch historiography still lacks an analysis of the relationship between infrastructural and transport developments, on the one hand, and economic development and industrialisation on the other. Only in one sector (Nusteling on the Rhine navigation) and one region (North Brabant by Passchier and Knippenberg on railway construction) have attempts been made. Important general questions concern the backward and forward effects of transport development. Around 1850 was the Dutch economy in need of a better transport system or not, and were transport innovations or improvements followed by measurable new economic activities and by regional specialisation of production? To answer these kinds of questions it will be necessary to study the specific features of transport developments, for instance volumes of traded goods, traffic flows, rate policies and specific interests of investors and town governments. It will also be necessary to distinguish between the types of improvement that were implemented — for instance, regarding speed, prices, reliability and freedom from seasonality.

As far as relations between the various transport modes are concerned. I suggest the following conclusions. Concerning the coming of the railways, it is necessary above all to draw a distinction between the carriage of passengers on the one hand and of goods on the other. The first railways in the Netherlands were built to meet the existing demand for passenger transport between the large cities in the province of Holland and brought about a revolution in speed. The first line was constructed only nine years after the first passenger railway: the Liverpool & Manchester of 1830, which proved innovative both in the use of steam locomotives and in its arrangements for the carriage of passengers and goods. By about 1870 the Dutch railways had practically monopolised the inland transport of passengers throughout the country. Although there was also a demand from industrialists for the transport of goods by railway, it was not strong enough to realise a national railway system before 1860. However, around 1860 the railways were a serious competitor to the Rhine navigation.

In general it seems that the existing waterways and roads were able to serve the existing demand for the inland transport of goods. Because of the mainly locally or regionally oriented economy there was only a little through traffic, apart from the movement of peat and navigation on the rivers. As far as we can speak of waterway networks, in the sense of well connected transport systems, their quality was poor, at least until 1880. The limitations of this system did not hamper or frustrate the economic process of industrialisation in the country. I conclude, on the basis of the literature, that infrastructure was not a bottleneck in economic development.

Developments in international traffic and transport and increasing inland industrial demand stimulated the construction of railways after 1860. Despite the fact that around 1885 the Netherlands had at its disposal an adequate railway system which was able to integrate the Dutch economy and which had improved connections with foreign countries, this system did not play a leading role in the inland movement of goods. This modest position of the railways was mainly due to the extension and integration of the waterways from 1865 onwards, for instance the opening up of the east and north of the country and the very strong position of navigation on the rivers.

Notes

1 R. T. Griffiths, *Industrial Retardation in the Netherlands 1830–1850* (The Hague, 1979), p. 66. See also R. W. J. M. Bos, 'Factorprijzen, technologie en marktstructuur: de groei van de Nederlandse volkshuishouding 1815–1914'. *A.A.G. Bijdragen*, 22 (1979), p. 125; *id.*, 'Van periferie naar centrum; enige kanttekeningen bij de Nederlandse industriële ontwikkeling in de negentiende eeuw', in P. A. M. Geurts and F. A. M. Messing (eds), *Economische ontwikkeling en sociale emancipatie*, II (The Hague, 1977). pp. 81–2; J. A. de Jonge, 'The role of the outer provinces in the process of Dutch economic growth in the nineteenth century', in Geurts and Messing, *Economische ontwikkeling*, pp. 57–9; C. Kreeft, 'Economische groei in Nederland', *Economisch- en Sociaal Historisch Jaarboek*, 51 (1988), pp. 226–9. For a brief survey of the historiography of the period 1813–1914 see J. L. van Zanden, 'The Dutch economic history of the period 1500–1940: a review of the present state of affairs', *Economic and Social History in the Netherlands*, I (1989), pp. 17–21.

2 The extent of railway per 100 km² in Great Britain, Belgium and the Netherlands in 1850 was respectively 4·0, 2·8 and 0·5 km. The figures are from B. R. Mitchell, *European Historical Statistics 1750–1970*, (1975), pp. 581, 582.

3 Griffiths, *Industrial Retardation*, p. 66.

4 H. C. Kuiler, *Verkeer en vervoer in Nederland. Schets eener ontwikkeling sinds 1815* (Utrecht, 1949), p. 2.

5 J. de Vries, *Barges and Capitalism: Passenger Transportation in the Dutch Economy 1632–1839* (Utrecht, 1981).

6 *Ibid.*, p. 222.

7 P. van Schaik, 'De economische betekenis van de turfwinning in Nederland', *Economisch-Historisch Jaarboek*, 32 (1969), pp. 191–200.

8 J. M. Fuchs, *Beurt- en wagenveren* (Amsterdam, 1946).

9 De Vries, *Barges*, pp. 53–7.

10 H. P. H. Nusteling, *De Rijnvaart in het tijdperk van stoom en steenkool 1831–1914. Een studie van het goederenverkeer en de verkeerspolitiek in de Rijndelta en het achterland, mede in verband met de opkomst van de spoorwegen en de concurrentie van vreemde zeehavens* (Amsterdam 1974), p. 220.

11 A. van de Woud, *Het lege land. De ruimtelijke orde van Nederland 1798–1848* (Amsterdam, 1987), pp. 118, 134.

12 Nusteling, *Rijnvaart*, p. 196. It was also very difficult for larger steamships to negotiate this stretch of waterway. The same goes for the Noord-Hollands canal, which was very narrow.

13 Kuiler, *Verkeer*, pp. 3–5.

14 Similar circumstances prevailed, for instance, in the relation between the provinces of Friesland and Drenthe. Van der Woud, *Het lege land*, pp. 121–5.

15 G. J. Schutten, *Varen waar geen water is* (Hengelo, 1981).

16 J. A. P. G. Boot, *De Twentsche katoennijverheid 1830–1873* (Amsterdam, 1935), p. 176.

17 In 1839 about 18,000 tons of freignt went from Almelo to Zwolle by road, and about 11,000 tons by ship. Schutten, *Varen*, pp. 127–8.

18 E. J. Fischer, *Fabriqueurs en fabrikanten. De Twentshe katoennijverheid en de onderneming. S.J. Spanjaard te Borne tussen circa 1800 en 1930* (Utrecht, 1983), pp. 50–1.

19 R. T. Griffiths, 'The creation of a national Dutch economy 1795–1909', *Tijdschrift voor Geschiedenis*, 95 (1982), p. 519; Kuiler, *Verkeer*, pp. 73–4. For instance, the following canals were built: in the north and the east the Oranjekanaal, Noordwillemskanaal, Overijssels Kanaal and extensions to the Hoogeveense vaart and Dedemsvaart (for the exploitation of peat deposits). In the south the Peelvaart, Noordervaart and Helenavaart.

Length of new canals opened in the Netherlands (km)

Period	km	Period	km
1811–20		1861–70	78
1821–30	250	1871–80	144
1831–40	36	1881–90	58
1841–50	93	1891–1900	95
1851–60	232	1901–1910	21

Source: Griffiths, 'Creation', p. 518.

20 Boot, *Twentsche*, p. 179.

21 Fischer, *Fabriqueurs*, p. 105, and Griffiths, *Industrial Retardation*, p. 77.

22 Boot, *Twentsche*, p. 183.

23 J. A. de Jonge, *De industrialisatie in Nederland tussen 1850 en 1914* (Nijmegen, 1976), p. 99.

24 C. C. M. Jacobs, 'De verkeerssituatie in de negentiende eeuw', in H. F. J. M. van den Eerenbeemt and H. J. A. M. Schurink (eds), *De opkomst van Tilburg als industrie-stad. Anderhalve eeuw economische en sociale ontwikkeling* (Nijmegen, 1959), p. 81. For an extensive analysis of the effects of the railways on the industrial development of North Brabant see N. Passchier and H. Knippenberg, 'Spoorwegen en industrialisatie in Nederland', *K.N.A.G. Geografisch Tijdschrift*, XII (1978), pp. 381–95.

25 On the economic impact of the building of the North Sea canal see T. van Tijn, 'Het Noordhollandse zeehavengebied, voor en na de openstelling van het Noordzeekanaal', *Tijdschrift voor Geschiedenis*, 79 (1966), pp. 279–319.

26 J. F. E. Bläsing, *Das goldene Delta und sein eisernes Hinterland 1815–1851. Von niederländisch-preussischen zu deutch-niederländischen Wirtschaftsbeziehungen* (Leiden, 1971). Nusteling, *Rijnvaart*. See note 2.

27 Nusteling, *Rijnvaart*, p. 130.

28 A history of these monopolies from the legal point of view is given by Fuchs, *Beurt- en Wagenveren*.

29 Kuiler, *Verkeer*, p. 51, and Griffiths, *Industrial Retardation*, p. 68. In the 1830s a trip from Rotterdam to Cologne by sailing boat would have taken ten to twelve days. In 1843 the same journey by steamship (towing iron barges) took about four or five days. See M. Hartgerink-Koomans, 'Amsterdam, Rotterdam en de Rijnspoor 1830–1856', *Bijdragen voor de Geschiedenis der Nederlanden*, 13 (1959), p. 28.

30 Nusteling, *Rijnvaart*, p. 219; Van der Woud, *Het lege iand*, pp. 112, 133.

31 *Ibid.*, p. 222. For a comprehensive treatment of the control of the rivers see *ibid.*, chapter 5.

32 Nusteling, *Rijnvaart*, p. 222.

33 G. P. van de Ven, 'Verbetering van de Rijntakken tussen 1850–1920', *Economisch- en Sociaal Historisch Jaarboek*, 46 (1983), p. 51.

34 *Ibid.*, pp. 229–43.

35 H. Schmal, ' 's Rijks groote wegen in de negentiende eeuw', in A. P. de Klerk (ed.), *Historische geografie in meervoud* (Utrecht, 1984), pp. 34–41.

36 De Vries, *Barges*, p. 359, n. 18.

37 These high roads had already been built in the seventeenth and eighteenth, centuries. See Schmal, ' 's Rijks groote wegen', p. 79.

38 Van der Woud, Het lege land, p. 142.

39 An exception was the huge *hessenwagens*, big carts, pulled by four or more horses, that could carry 2·5 tons of freight.

40 De Vries, *Barges*, pp. 60–1.

41 *Ibid.*, p. 166.

42 *Ibid.*, p. 169.

43 For instance, in 1847 the coach service between 's-Gravenhage and Rotterdam reached a peak, with twenty-four departures a day. In the same year the railway connection between the two cities was completed. See Schmal, ' 's Rijks', p. 84.

44 *150 jaren Van Gend & Loos 1796–1946* (Utrecht, 1946), p. 44.

45 *Ibid.*, pp. 86–7.

46 Griffiths, 'Creation', p. 518.

47 In the period 1841–50 a yearly average of 28,000 carts passed the town of Nijverdal on the new road from Almelo to Zwolle. In 1847 35,000 carts arrived at Enschede, carrying a total of 47,424 tons. See Schutten, *Varen*, p. 23.

48 *Ibid.*, p. 24.

49 Schmal, ' 's Rijks', p. 88.

50 J. H. Jonckers Nieboer, *Geschiedenis der Nederlandsche spoorwegen* (Rotterdam, 1938).

51 See notes 1 and 5.

52 The only exception is the estimate by J. de Vries (*Barges*, p. 207) of the social savings of the railways in the movement of passengers in the period 1848–53.

53 W. van den Broeke, *Financiën and financiers van de Nederlandse spoorwegen 1837–1890* (Zwolle, 1985).

54 Jonckers Nieboer, *Geschiedenis*, chapter 1.

55 *Ibid.*, pp. 17–19.

56 Hartgerink-Koomans, 'Amsterdam', p. 53.

57 G. A. van der Knaap, *A Spatial Analysis of the Evolution of an Urban System: the Case of the Netherlands* (Rotterdam, 1978), pp. 76–8.

58 For instance, in the building of railway goods yards provision for freight traffic such as storage facilities was not taken into account. See *Gendenkboek der Hollandsche IJzeren Spoorweg-Maatschappij 1839–1889* (Amsterdam, 1889), pp. 73–5. According to Griffiths this was simply a matter of costs; the high capital and maintenance costs raised the cost of freight above that of the much cheaper waterways. See Griffiths, *Industrial Retardation*, p. 74.

59 De Vries, *Barges*, p. 206.

60 *Gedenkboek*, p. Q: Overzicht van de ontwikkeling van het verkeer op de lijn Amsterdam–Rotterdam tevens in verband met den loop der bevolking (table illustrating the growth of traffic on the Amsterdam–Rotterdam railway and the increase in the population 1839–88).

61 Van der Woud, *Het lege land*, p. 194.

62 There are several explanations for this delay. Nusteling states that Germany agreed only after obtaining free passage down the Rhine to the sea. See Nusteling, *Rijnvaart*, p. 104. Jonckers Nieboer ascribes it to difference in gauge; that the Dutch railways was 2 m, against the standard 1·435 m. See Jonckers Nieboer, *Geschiedenis*, p. 53. See also S. Boom and P. Saal, 'Spoorwegaanleg en het beeld van de eerste helft van de negentiende eeuw', *Economisch- en Sociaal Historisch Jaarboek*, 46 (1983), pp. 18–19.

63 Nusteling, *Rijnvaart*, pp. 103–4.

64 M. Hartgerink-Koomans, 'Handelsbetrekkingen en spoorwegverbindingen in de eerste helft der negentiende eeuw. Plannen tot spoorwegbouw in de noord-oostelijke provincies 1845–1854', *Economisch Historisch Jaarboek*, 26 (1956), p. 47. See also R. Dijksterhuis, *Spoorwegtracering en stedebouw in Nederland. Historische analyse van een wisselwerking* (Delft, 1984), pp. 13, 26.

65 M. Hartgerink-Koomans, 'De Zuiderlijnen. Opzet en bestrijding der spoorlijnen beneden de grote rivieren, *Economisch Historisch Jaarboek*, 30 (1965), pp. 11, 58.

66 Griffiths, 'Creation', p. 519.

67 Passchier and Knippenberg, 'Spoorwegen', p. 394.

68 In 1880 the railway companies earned 11·6 million guilders from passenger traffic and 9·6 million from goods. In 1900 the figures were 20·7 and 19·9 respectively. Nusteling, *Rijnvaart*, p. 495.

69 H. C. Kuiler, *Verkeer en vervoer in Nederland, zoals deze in hun recente ontwikkeling zijn bepaald door economisch-geografische factoren* (Utrecht, 1946), p. 45.

70 See note 2. This book contains an English and German summary.

71 Kuiler, *Verkeer*, p. 76; Nusteling, *Rijnvaart*, pp. 111–16.

72 In the case of coal traffic from Germany to Amsterdam, however, the railways were in a favourable position. The river Ruhr was hardly navigable for large coal vessels, and railways had to be used to carry the coal from the mines to the Rhine harbours. Furthermore, with differential rates set very high for short hauls, the railways encouraged consignment direct to the Netherlands. See, for instance, Nusteling, *Rijnvaart*, pp. 176–9.

73 In 1879 Prussia had nationalised the private railway companies. In order to promote the transport of goods by railway to and from German seaports, a reduction in railway rates on through traffic via the Dutch seaports, to compete with river navigation, was blocked. In the years 1878–80 the yearly average of goods sent by rail to and from Germany amounted to 2 million tons; in 1894–96 the figure was 4·5 million. Rhine shipments to and from Germany totalled 3·4 million and 8·3 million tons annually, for the same periods respectively. After this period the difference between carriage by rail and by Rhine navigation became much greater. Nusteling, *Rijnvaart*, p. 357.

74 *Ibid.*, p. 357.

75 De Vries, *Barges*, p. 215.

76 W. Fritschy, 'Van trekvaarttechnologie naar spoorwegtechnologie', *Bijdragen en Mededelingen Betreffende de Geschiedenis der Nederlanden*, 95 (1980), p. 521.

77 The dividends of the HSM varied in the 1840s between 3 and 4 per cent, while the average value of the shares was below par between 1840 and 1862, except for the years 1844 and 1845. See Jonckers Nieboer, *Geschiedenis*, appendix K.

78 *Ibid.*, p. 71.

79 Van der Woud, *Het lege land*, pp. 190–2.

80 See note 50.

81 See, for instance, A. Fishlow, *American Railroads and the Transformation of the Ante-Bellum Economy* (Cambridge, Mass., 1965); R. Fogel, *Railroads and American Economic Growth* (Baltimore, 1964); W. W. Rostow, 'Leading sectors and the take-off', in W. W. Rostow (ed.), *The Economics of Take-off into Sustained Growth* (New York, 1963), pp. 1–21.

82 For instance, in the Twente region the two brothers Salomonson, cotton manufacturers, tried to finance a railway between Arnhem and northern Germany. These plans for a 'Overijsselsche Spoorwegmaatschappij' failed. See Jonckers Nieboer, *Geschiedenis*, p. 40. Other industrialists were involved with railway plans. See, for example, W. Fritschy, 'Spoorwegaanleg in Nederland van 1831 tot 1845 en de rol van de Staat daarin', *Economisch- en Sociaal Historisch Jaarboek*, 45 (1983), p. 195; Hartgerink-Koomans, 'Handelsbetrekkingen' and 'Zuiderlijnen'.

83 Van den Broeke, *Financiën*, pp. 158–9.

84 Boom and Saal, 'Spoorwegaanleg', p. 6.

85 Bos, 'Periferie', p. 88.

Acknowledgement

I would like to thank Rainer Fremdling and an anonymous referee for helpful suggestions and comments.